Lecture Notes in Artificial Intelligence   4651

Edited by J. G. Carbonell and J. Siekmann

Subseries of Lecture Notes in Computer Science

Francisco Azevedo   Pedro Barahona
François Fages   Francesca Rossi (Eds.)

Recent Advances in Constraints

11th Annual ERCIM International Workshop
on Constraint Solving and
Contraint Logic Programming, CSCLP 2006
Caparica, Portugal, June 26-28, 2006
Revised Selected and Invited Papers

 Springer

Series Editors

Jaime G. Carbonell, Carnegie Mellon University, Pittsburgh, PA, USA
Jörg Siekmann, University of Saarland, Saarbrücken, Germany

Volume Editors

Francisco Azevedo
Universidade Nova de Lisboa
2829-516 Caparica, Portugal
E-mail: fa@di.fct.unl.pt

Pedro Barahona
Universidade Nova de Lisboa
2825 Caparica, Portugal
E-mail: pb@di.fct.unl.pt

François Fages
INRIA Rocquencourt - Projet CONTRAINTES
BP 105, 78153 Le Chesnay Cedex, France
E-mail: Francois.Fages@inria.fr

Francesca Rossi
University of Padova
35131 Padova, Italy
E-mail: frossi@math.unipd.it

Library of Congress Control Number: 2007931597

CR Subject Classification (1998): I.2.3, F.3.1-2, F.4.1, D.3.3, F.2.2, G.1.6, I.2.8

LNCS Sublibrary: SL 7 – Artificial Intelligence

ISSN   0302-9743
ISBN-10   3-540-73816-9 Springer Berlin Heidelberg New York
ISBN-13   978-3-540-73816-9 Springer Berlin Heidelberg New York

Springer is a part of Springer Science+Business Media

springer.com

© Springer-Verlag Berlin Heidelberg 2007
Printed in Germany

Typesetting: Camera-ready by author, data conversion by Scientific Publishing Services, Chennai, India
Printed on acid-free paper   SPIN: 12096343   06/3180   5 4 3 2 1 0

Preface

Constraints are a natural way to represent knowledge, and constraint programming is a declarative programming paradigm successfully used to solve many difficult combinatorial problems. Examples of application domains where such problems naturally arise, and where constraint programming has made a valuable contribution, are scheduling, production planning, communication networks, robotics, and bioinformatics.

This volume contains the extended and reviewed versions of a selection of papers presented at the 11th International Workshop on Constraint Solving and Constraint Logic Programming (CSCLP 2006), that was held during June 26–28, 2006 at the New University of Lisbon, Portugal. It also contains papers that were submitted in response to the open call that followed the workshop. Both types of papers were reviewed independently by three experts in the specific topics.

The papers in this volume present original research results, as well as applications, in many aspects of constraint solving and constraint logic programming. Research topics that can be found in the papers are symmetry breaking, privacy, distributed forward checking, quantified CSPs, bipolar preferences, first-order constraints, microstructure, constraint handling rules, acyclic clustered problems, as well as the analysis of application domains such as disjunctive resource problems and stochastic inventory control. Moreover, the volume also contains a tutorial on hybrid algorithms.

The editors would like to take the opportunity to thank all the authors who submitted a paper to this volume, as well as the reviewers for their helpful work. This volume was made possible thanks to the support of the European Research Consortium for Informatics and Mathematics (ERCIM), the Association for Constraint Programming (ACP), the Portuguese Foundation for Science and Technology (FCT), the Department of Computer Science of the New University of Lisbon, and its Centre for Artificial Intelligence (CENTRIA).

We hope that the present volume is useful to everyone interested in the recent advances and new trends in constraint programming, constraint solving, problem modelling, and applications.

March 2007

F. Azevedo
P. Barahona
F. Fages
F. Rossi

Organization

CSCLP 2006 was organized by the ERCIM Working Group on Constraints.

Organizing and Program Committee

Francisco Azevedo CENTRIA, FCT/UNL, Portugal
Pedro Barahona   CENTRIA, FCT/UNL, Portugal
François Fages   INRIA Rocquencourt, France
Francesca Rossi   University of Padova, Italy

Additional Reviewers

R. Barták	N. Jussien	B. Peintner
M. Basharu	A. Kovacs	S. Prestwich
M. Benedetti	A. Lallouet	J.-F. Puget
T. Benoist	M. Maher	S. Soliman
D. Le Berre	F. Manyà	J. Vautard
S. Bistarelli	M. Meister	M. Wallace
K. Cheng	P. Meseguer	R. Wallace
K. Djelloul	M. Moffitt	R. Yap
T. Frühwirth	E. Monfroy	N. Yorke-Smith
B. Hnich	B. O'Sullivan	P. Zoeteweij

Sponsoring Institutions

ERCIM Working Group on Constraints
Association for Constraint Programming
Universidade Nova de Lisboa
Centro de Inteligência Artificial, Portugal
Fundação para a Ciência e Tecnologia, Portugal

Table of Contents

Hybrid Algorithms in Constraint Programming

Mark Wallace

Monash University, Faculty of Information Technology,
Building 63, Clayton, Vic. 3800, Australia
mark.wallace@infotech.monash.edu.au
http://www.infotech.monash.edu.au/

Abstract. This paper surveys hybrid algorithms from a constraint programming perspective. It introduces techniques used within a constructive search framework, such as propagation and linear relaxation, as well as techniques used in combination with search by repair.

Keywords: constraint programming, hybrid algorithms, search.

1 Introduction

1.1 Tribes

There are three research communities exploring combinatorial optimisation problems. Within each community there is strong debate and ideas are shared naturally. Between the communities, however, there is a lack of common background and limited cross-fertilisation.

We belong to one of those communities: the CP community.[1] The other two are Mathematical Programming (MP) and Local Search and meta-heuristics (LS). Currently LS seems to be the largest of the three. It has become clear that such a separation hampers progress towards our common goal, and there should be one larger community - whose name is a point of contention - which should include us all.

Hybrid algorithms lie at the boundary between CP, MP and LS. We will explore some of the techniques used in MP and LS, and show how they can be used in conjunction with CP techniques to build better algorithms. We will not here be investigating the "frontiers of research" in these communities. However it is my belief that CP can contribute right now at these frontiers. Hybrid techniques are not peripheral to the research of any of these communities. They are the key to real progress in all three.

1.2 Overview

Firstly we explore the mapping of problems to algorithms, the requirement for problem decomposition, and the need for linking solvers and solver cooperation.

---

[1] There are also, of course, many people in the CP community who are not exploring combinatorial optimisation problems.

F. Azevedo et al. (Eds.): CSCLP 2006, LNAI 4651, pp. 1–32, 2007.

Different ways of linking solvers will be discussed, and some of their benefits and applications.

Secondly we will investigate different kinds of search, coming from the different communities, and see how they can be used separately, and together.

The paper is presented from a CP viewpoint, aimed at a CP audience. However, the objective is to lower the barrier to exploiting hybrid techniques, encouraging research at the boundaries of CP, MP and LS, and finally to help bring these communities together.

2 Hybrid Constraint Solving

2.1 The Conceptual Model and the Design Model

To solve a problem we start with a problem-oriented *conceptual* model. The syntax of conceptual models is targeted to clarity, expressive power and ease of use for people interested in solving the problem.

The conceptual model is mapped down to a *design* model which is machine-oriented [Ger01]. The design model specifies the algorithm(s) which will be used to solve the problem at a level that can be interpreted by currently implemented programming platforms, like ECLiPSe [AW06].

The CP community usually separates the *model* from the *search strategy*. The design model in our terminology includes both a model with variables and constraints and a search strategy. The variables and constraints in the design model are chosen so as to be easy to solve, and to fit with the search routine. Moreover with each constraint in the design model solving methods are specified which associate a behaviour with the constraint. Though the variables and constraints in the design model of a problem may be quite different from those in its conceptual model, they are logically equivalent - they represent the same set of solutions. In principle, the conceptual modeling language could be a subset of the design modeling language.

Real problems are complex and, especially, they involve different kinds of constraints and variables. For example a "workforce scheduling" problem [AAH95] typically involves the following decision variables:

- For each task, one or more *employees* assigned to the task.
- For each task, a *start time*
- For each (group of) employee(s), a *route* that takes them from task to task.
- For each (group of) employee(s), *shift start, end, and break times*

This is in sharp contrast to typical CSP puzzles and benchmarks, such as graph colouring, where all the variables are of the same "kind" and sometimes even have the same initial domains.

The constraints and data in real problems are also diverse. The workforce scheduling problem includes:

- *Location and distance* constraints on and between tasks
- *Skills* data and constraints on and between employees and tasks
- *Time* constraints on tasks and employee shifts

Naturally there are many more constraints in real workforce scheduling problems on vehicles, road speeds, equipment, team composition and so on.

The consequence is that the algorithm(s) needed to solve real problems are typically *hybrid*. The skills constraints are best solved by a different sub-algorithm from the routing constraints, for example.

2.2   Mapping from the Conceptual to the Design Model

To map a problem description to an algorithm, it is often necessary to decompose the whole problem into parts that can be efficiently solved. The challenge is to be able to glue the subproblem solutions together into a consistent solution to the whole problem. Moreover, for optimisation problems, it is not enough to find the optimal solution to each subproblem. Glueing these "local" optima together does not necessarily yield a "global" optimum.

For these reasons we need to ensure that the subproblem algorithms cooperate with each other so as to produce solutions that are both consistent with each other and, as near optimal as possible. The design of *hybrid algorithms* that meet these criteria is the topic of this section.

In principle we can map a conceptual model to a design model by

- Associating a behaviour, or a constraint solver, with each problem constraint
- Adding a search algorithm to make up for the incompleteness of the constraint solvers

In practice the design model produced by any such mapping is strongly influenced by the particular choice of conceptual model. The "wrong" conceptual model could make it very hard to produce an efficient algorithm to solve the problem.

For this reason we must map a given conceptual model to an efficient design model in two steps:

- Transform the conceptual model into another one that is more suitable for mapping
- Add constraint behaviour and search, to yield an efficient design model

The first step - transforming the conceptual model - is an art rather than a science. It involves four kinds of transformations:

- Decomposition - separating the constraints and variables into subproblems
- Transformation - rewriting the constraints and variables into a form more suitable for certain kinds of solving and search
- Tightening - the addition of new constraints whose behaviour will enhance efficient problem solving
- Linking - the addition of constraints and variables that will keep the separate subproblem solvers "in step" during problem solving

The decomposition is germane to our concerns. It is therefore worth discussing briefly here. Firstly, we note that the decomposition covers the original problem (of course), but it is *not* a partition: otherwise the subproblems would have no link whatsoever between them.

Therefore some subproblems share some variables.[2] Each subproblem solver can then make changes to a shared variable, which can be used by the other solver. Sometimes constraints are shared by different subproblems. In this case the same constraint is handled multiple times by different solvers, possibly yielding different and complementary information within each solver. When the constraints in different subproblems are transformed in different ways, the result is that the same constraint may appear several times in several different forms in the transformed conceptual model. We shall see later a model with a resource constraint that is written three different ways for three different solvers.

We shall now move on to examine design models for a variety of hybrid algorithms.

3   Constraint Solvers

In this section we discuss different constraint solvers, and constraint behaviours. We investigate what kinds of information can be passed between them, in different hybrid algorithms, and how their cooperative behaviour can be controlled.

The solvers, and constraint behaviours, we will cover are

- Finite domain (FD) constraint propagation and solving
- Global constraints and their behaviour
- Interval constraints, and bounds propagation
- Linear constraint solving
- Propositional clause (SAT) solving
- Set constraint solving
- One-way constraints (or "invariants")

Referring back to the three research communities, we can relate these solvers to the CP and MP communities. Accordingly this work lies on the border of CP and MP. The hybrids on the border of CP and LS will be explored in the next section.

3.1   Modelling Requirements for Constraint Solvers

Supposing a constraint C appears in the conceptual model and it is to be handled by a particular solver, say FD. Then the design model must express a number of requirements.

Firstly it must associate an initial domain which each of the variables in em C. For the solver FD, the initial domains are discrete, and finite. For other solvers,

---

[2] This is a simplification. For example mathematical decomposition techniques such as Lagrangian relaxation and column generation use more sophisticated techniques than shared variables to relate the subproblems and the master problem. These will be discussed in sections 4.2 and 5.1 below.

such as SAT or linear, they must be initialised as booleans, or reals as required by the solvers.

Secondly the constraint should be explicitly associated with a particular solver. It may also be associated with more than one solver - and this is the subject of section 4 below.

Thirdly its activation conditions must be specified. A constraint can be activated by simply sending it to the solver, so that it is automatically handled whenever the solver is invoked, or by explicitly introducing conditions under which it should be woken. For an almost linear constraint, for example, which includes a few non-linear expressions, the constraint might only be woken and sent to the linear solver once all its expressions have become linear. The most common use of explicit waking is for propagation constraints, discussed in the following subsection.

No further specification about the constraints themselves need be given in the design model. Details about what information is passed to the constraint before solving it, and what information is extraced from the constraint after solving it is a property of the solver, rather than the individual constraint.

In principle, when there is a single search routine with a current state, the solvers communicate two key types of information to the state:

1. Satisfiability or inconsistency
2. Variable values

Other information can be extracted from the different solvers by explicit requests expressed in the design model. Variable values are also, typically, passed from the search state to the solvers. Thus when one solver, say FD, instantiates a variable, this information is made available to all the other solvers.

3.2 Constraints Which Propagate Domain Information

Information Exported. We first consider finite domains and global FD constraints. The relevant issues for hybrid algorithms are

- what information can be extracted from the solver
- under what conditions *all* the information has been extracted from the solver: i.e. the extracted information entails the constraint, so can we be sure that a state which appears to be feasible for the other subproblems and their solvers is also guaranteed to be consistent with the FD solver.

The answers are as follows:

- Information that can be extracted from the solver includes upper and lower bounds on the variables, domain size and if necessary precise information about which values are in the domain of each variable. The solver also reports inconsistency whenever a domain becomes empty.
- All the information has been extracted from a constraint when it is entailed by the current (visible) domain information. An FD solver can sometimes

detect when this is the case and "kill" the constraint. Until then, the constraint is still "active". Therefore active constraints are ones which, to the FD solver's knowledge, are not yet entailed by the domains of the variables. Some FD solvers don't guarantee to detect this entailment until all the variables have been instantiated. Many FD solvers support *reified* constraints, which have a boolean variable to flag entailment or disentailment (inconsistency with the current variable domains).

The domain, inconsistency and entailment information are all logical consequences of the FD constraints and input variable domains. For this reason, no matter what other constraints in other subproblems are imposed on the variables, this information is still correct. In any solution to the whole problem, the values of the FD variables must belong to the propagated domains. Inconsistency of the subproblem, implies inconsistency of the whole problem. If the variable domains entail the subproblem constraints, then they are still entailed when the constraints from the rest of the problem are considered.

Global Constraints. Notice that global constraints are often themselves implemented by hybrid techniques, even though the information imported and exported is restricted to the above. For example a feasible assignment may be recorded internally, so as to support a quick consistency check, or to speed up the propagation algorithm. An interesting case of hybridisation is the use of continuous variables in global constraints. The classic example is a global constraint for scheduling, where resource variables are FD, but the task start time variables could be continuous. As far as I know the hybrid discrete/continuous scheduling constraint is not yet available in any CP system.[3]

Interval Constraints. For interval constraint solvers only upper and lower bounds, and constraint entailment are accessible. The problem with continuous constraints is that they are not necessarily instantiated during search. Since continuous variables can take infinitely many different values, search methods that try instantiating variables to all their different possible values don't necessarily terminate. Instead search methods for continuous values can only tighten the variable's bounds, until the remaining interval associated with the variable becomes "sufficiently" small.

Not all values within these small intervals are guaranteed to satisfy all the constraints. Indeed there are common cases where, actually, there are no feasible solutions, even though the final intervals appear prima facie compatible. One vivid example is Wilkinson's problem (quoted in [Van98]). It has two constraints: $\prod_{i=1}^{20}(X+i) + P \times X^{19} = 0$ and $X \in [-20.4.. -9.4]$. When $P = 0$ the constraints have 11 solutions ($X = -10 \ldots X = -20$), but when P differs infinitesimally from 0 (viz. $P = 2^{-23}$), it has no solutions!

For these reasons "answers" returned by search routines which associate small intervals with continuous variables are typically conjoined with a set of undecided constraints, which must be satisfied in any solution.

---

[3] CP scheduling will be covered in more detail later.

3.3   Linear Constraints

Underlying Principles. A linear constraint solver can only handle a very restricted set of constraints. These are linear numeric constraints that can be expressed in the form $Expr \geq Number$ or $Expr \leq Number$. The expression on the left hand side is a sum of *linear terms*, which take the form *Coefficient* $\times$ *Variable*. The coefficients, and the number on the right hand side, are either integers or real numbers [Wil99].

Linear constraint solvers are designed not only to find feasible solutions, but also to optimise against a cost function in the form of another linear expression.

In the examples in this chapter we shall typically write the linear constraints in the form $Expr \geq Number$, and assume that the optimisation direction is *minimisation*.

Whilst much less expressive than CP constraints, they have a very important property: any set of linear constraints, over real variables, can be tested for *global* consistency in polynomial time. This means we can throw *all* the linear constraint of a problem into the linear solver and immediately determine whether they are consistent.

By adding just one more kind of constraint, an *integrality* constraint that requires a variable to take only integer values, we can now express any problem in the class *NP*. (Of course the consistency problem for mixed linear and integrality constraints - termed MIP, for "Mixed Integer Programming" - is NP-hard).

The primary information returned by the linear solver is consistency or inconsistency among the set of linear constraints. However for building cooperative solvers we will seek more than this.

Firstly the solver can also export an optimal solution - assuming throughout this section that the cost function is linear. In general there may be many optimal solutions, but even from a single optimum we now have a known optimal value for the cost function. No matter how many other constraints there may be in other solvers, the optimal value cannot improve when they are considered, it can only get worse. Thus the linear solver returns a bound on the cost function.

Linear constraints are special because if S_1 and S_2 are two solutions (two complete assignment that satisfy all the linear constraints), then any assignment that lies on the line between S_1 and S_2 is also feasible. For example if $X = 1$, $Y = 1$ is a solution, and so is $X = 4$ and $Y = 7$, then we can be sure that $X = 2$ and $Y = 3$ is a solution, and so is $X = 3$ and $Y = 5$. Moreover since the cost function is linear, the cost of any solution on the line between $S1$ and $S2$ has a cost between the cost of $S1$ and the cost of $S2$.

These properties have some important consequences. Supposing $Expr \geq Number$ is a constraint, and that at the optimal solution the value of $Expr$ is strictly greater than $Number$. Then the problem has the same optimal value even if this constraint is dropped (or "relaxed"). Otherwise you could draw a line between a new optimal solution and the old one, on which all points are feasible for the relaxed problem. Moreover the cost must decrease continuously towards the new optimum solution. Therefore at the point where this line crosses the line $Expr = Number$ (i.e. at the first point where the solution is also feasible for the original problem) the cost is less than at the old optimal solution, contradicting the optimality of the original solution.

In short, for a linear problem all the constraints which influence the optimal cost are *binding* at an optimal solution, in the sense that the expression on the left hand side is equal to the number on the right.

Shadow Prices and Dual Values. If such a binding constraint was dropped, then the relaxed problem typically would have a new optimum value (with a solution that would have violated the constraint). Along the line between the old and new optimum, the cost function steadily improves. Instead of relaxing the constraint we can decrease the number on its right hand side so as to partially relax the constraint. As long as the constraint is still binding, at the new optimum the expression on the left hand side is equal to the new number on the right hand side. We can measure how much the optimum value of the cost function improves as we change the number on the right hand side of the constraint. This ratio is called the "shadow price" of the constraint.[4]

Indeed using the shadow price we can relax the constraint altogether, and effectively duplicate it in the optimisation function. If λ is the shadow price then dropping $Expr \geq Number$ and adding $\lambda \times (Number - Expr)$ to the cost function, we have a new problem with the same optimum solution as the old one.

There is another very interesting way to approach the very same thing. If we have a set of linear constraints of the form $Expr \geq Number$, we multiply each constraint by a positive number and we add all the constraints together, i.e. we add all the multiplied left hand sides to create a new expression $SumExpr$ and we add all the multiplied right hand sides to get a new number $SumNumber$, then $SumExpr \geq SumNumber$ is again a linear constraint. Surprisingly everything you can infer from a set of linear constraints you can get by simply performing this one manipulation: forming a *linear combination* of the constraints.

In particular if you multiply all the constraints by their shadow prices at the optimum, and add them together the final right hand side is the optimum cost, and the left hand side is an expression whose value is guaranteed to be smaller than the expression defining the cost function. Indeed of all the linear combinations of the constraints whose left hand side is dominated by the cost function, the one formed using the shadow prices has the highest value on the right hand side.

We call the multipliers that we use in forming linear combinations of the constraints *dual values*. Fixing the dual values to maximize the right-hand-side expression $SumNumber$ is a way of finding the optimal cost. At the optimal solution, the dual values are the shadow prices.

Simplex and Reduced Costs. In the section on Underlying Principles, above, we showed that at an optimal solution the set of tight constraints are the only

---

[4] Technically the shadow price only takes into account those constraints which were binding at the current optimal solution. If at the new optimal solution for the relaxed problem another constraint became binding, then the shadow price would be an overestimate of the improvement in the optimum value.

ones that matter. In general if the problem has n decision variables, then it must have at least n linearly independent (i.e. different) constraints otherwise the cost is unbounded. Indeed there must be an optimal solution with n (linearly independent) tight constraints. Moreover n such constraints define a unique solution.

Consequently one way of finding an optimal solution is simply to keep choosing sets of n linearly independent constraints, making them tight (i.e. changing the inequality to an equality), and computing their solution until there are no more such sets. Then just record the best solution. In fact this can be made more efficient by always modifying the current set into a new set that yields a better solution.

We can rewrite an inequality constraint as an equation with an extra (positive) variable, called the "slack" variable. The n slack variables associated with the n tight constraints are set to zero. If there are m inequality constraints, the total number of variables (decision variables and slack variables) is $m + n$. The other m variables are constrained by m equations, and we say they are *basic* variables.

An optimal solution can be found by a hill climbing algorithm (the "Simplex" algorithm) which at each steps swaps one variable out of the basis and another in. The variables swapped into the basis is computed using a function called the *reduced cost*. For each current basis, a reduced cost can be computed for each non-basic variable, and any variable with a negative reduced cost has the potential to improve the solution. Another variable is chosen to be swapped out, and if the swap indeed yields a better solution, then the move is accepted. When no non-basic variable has a negative reduced cost the current solution is optimal.

Information Exported from the Linear Constraint Solver. The information that can be extracted from a linear constraint solver is as follows:

- An optimal bound on the cost variable, a shadow price for each constraint, and a reduced cost for each problem variable. Additionally one solution with optimal cost can be extracted from the solver.
- The solver reports inconsistency whenever the linear constraints are inconsistent with each other

Upper and lower bounds for a variable X can be elicited from a linear solver by using X and $-X$ in turn as the cost function. However these bounds are computationally expensive to compute, and even for linear constraints, the FD solver typically computes bounds much more cheaply.

Unlike the FD solver, there is no problem of consistency of the constraints inside the solver, but the problem remains how to ensure these are consistent with the constraints in other solvers.

3.4 Propositional Clause Solvers and Set Solvers

Proposition clause solvers are usually called SAT solvers. SAT solvers are typically designed to find as quickly as possible an instantiation of the propositional (i.e. boolean) variables that satisfies all the clauses (i.e. constraints). Many SAT solvers, such as zChaff [ZMMM01], generate nogoods, expressed as clauses.

The challenge, for hybridising a SAT solver with other solvers is to interrupt the SAT solving process - which in general includes an exponential-time search procedure - before it has necessarily found a solution, to extract useful information from it, and to return it to the same state as when the SAT solver was invoked.

Information that can be extracted from a SAT solver is as follows:

- Nogood clauses
- Feasible solutions (in case the solver terminates with success before being interrupted)
- Inconsistency (in case the solver terminates with failure before being interrupted)

Set solvers typically only handle finite sets of integers (or atoms). Their behaviour is similar to FD solvers, except they propagate upper and lower bounds (i.e. the set of element that *must* belong to the set, and the set of elements that *could* still belong to the set. The set cardinality is typically handled as an FD variable.

Information that can be extracted from a set solver is as follows:

- Upper and lower bounds
- Set cardinality
- The solver reports inconsistency whenever the upper bound ceases to be a superset of the lower bound

All the variables in a SAT solver are booleans. Set solvers also can be transformed into a representation in terms of boolean variables (associate a boolean variable with each value in the upper bound, setting those already in the lower bound to one, and setting the cardinality to the sum of the booleans).

3.5   One-Way Constraints, or Invariants

Informally, a one-way solver is simply a solver that evaluates a function each time any of its argument values are changed, and returns the new result. The implementatio challenge is to evaluate the function very quickly, using the previous result and just computing the impact of the changes.

Historically constraint programming is related to theorem proving, and constraint programs are often thought of as performing very fast logical inferences. Thus, for example, from the constraints $X = 2 + Y$ and $Y = 3$ the constraint program infers $X = 5$, which is a logical consequence.

Nevertheless there is information exported from constraint solvers which is very useful for guiding search but has no logical interpretation. One-way solvers can propagate and update such *heuristic* information very efficiently.

For many problems it is often useful to export a *tentative* value for a variable. This has no logical status, but is also very useful for guiding heuristics. This may be the value of the variable in a solution to a similar problem, or it may be the value of the variable in the optimal solution of a relaxed version of the problem.

In a decomposition where the variable belongs to more than one new subproblem, the tentative value may be its value in a solution of one subproblem.[5]

Now if variable Y has tentative value 3, and the (functional) constraint $X = 2 + Y$ is being handled by a one-way solver, then the tentative value of X will be set by the solver to 5. If the tentative value of Y is then changed, for some reason, to 8, then the tentative value of X will be updated to 10.

A one-way solver, then, computes the value of an expression, or function, and uses the result to update the (tentative) value of a variable. A one-way solver can also handle constraints by reifying them, and then updating the value of the boolean associated with the constraint. This enables the programmer to quickly detect which constraints are violated by a tentative assignment of values to variables.

If a variable becomes instantiated, the one-way solver treats this as its (new) tentative value, and propagates this to the tentative values of other variables as usual.

Information that can be extracted from a one-way solver is as follows:

- Tentative values for variables.
- Constraints violated by the tentative assignment

4   Communicating Solvers

We introduce this section with a diagram showing the standard types of information communicated between solvers. Obviously there are many forms of communication not shown in this diagram. For example if *SAT* represents set membership through boolean variables, then it can communicate with *Set*. However this would have to be explicitly introduced in the design model via additional *channeling* constraints, introduced below in section 4.1. Secondly the communication of bounds information illustrated in this diagram is redundant. As long as the em IC solver is there, it can forward bounds changes between *FD* and em linear. However not all design models include an *IC* solver, and in this case the direct communication between *FD* and *linear* is needed.

Constraint solvers can cooperate by sharing *lemmas* about the problem, or by sharing heuristic information. A lemma, from the CP viewpoint, is just a redundant constraint that can be used by the other solver to help focus and prune search. These lemmas can take many different forms: they may be *nogoods*, *cutting planes*, generated *rows*, *fixed variables*, tightened *domains*, propagated *constraints*, *cost bounds*, and even *reduced costs* [FLM04].

Some lemmas are logical consequences of the problem definition: these include global cuts and cost bounds. Other lemmas are valid only on the assumption that certain extra constraints have been imposed (during search). Validity in this case means that the problem definition conjoined with the specified extra constraints entail the lemma. A typical example is the information propagated

---

[5] Indeed a variable may have several tentative values, but we shall only consider a single tentative value here.

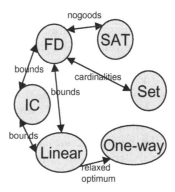

Fig. 1. Communicating Solvers

during search by a bounds consistency algorithm. Naturally this information is only true for those parts of the search space lying below the search tree node where the information was inferred.

Heuristic information can also be global or local. Global heuristic information includes the number of constraints on a variable, an optimal solution to the linear relaxation problem at the "root" node of the search tree. Local heuristic information might include the size of the variables domains at a certain point during search, or the shadow prices of the linear constraints at the current node in the branch and bound search tree.

The key difference between lemmas and heuristic information lies in the consequences of errors.

If the heuristic information is out of date, or wrong, then the algorithm performance may be affected but the solution produced by the algorithm remains correct. However if lemmas are wrong, then the algorithm will generally yield some wrong answers.

4.1 Channeling Constraints

The same problem often needs to be modelled using different variables and constraints for efficient handling in different solvers. The n-queens problem for example can be modelled by n variables each with domain $1..n$, where each variable represents a queen in a different row, and each value represents the column in which that queen is placed. This is ideal for FD solvers, and disequations. It can also be modelling by n^2 zero-one variables, one for each square on the board. This is best for integer linear constraints where a constraint that two queens cannot share a line is encoded by constraining the sum of all the boolean variables along the line to be less than or equal to one.

All CSP problems can be transformed to SAT problems, using the same mapping of FD variables to booleans. Some researchers hope to achieve such high performance for SAT solvers that this transformation will prove the best way to solve all CSP problems [AdVD+04].

SAT can perform very competitively on some typical puzzles and simple benchmarks. However the programmer has a great deal of control over FD search which is both a disadvantage (because SAT search is automatic) and an advantage because the programmer can tune the search to the problem at hand.

CSP problems can also be transformed to integer/linear problems, and some mathematical programmers believe that all such problems can be formulated (as a design model) in such a way that the integer/linear solver offers the fastest solution method. Whilst there is some truth in this claim, there are many problems and puzzles for which CP outperforms the best integer/linear models designed to date.

The point here is that all the different kinds of solvers - FD, interval, integer/linear, SAT, set - are sometimes more suitable and sometimes not so suitable as other solvers, *and to date we have discovered no way of inferring from the conceptual model which will be the best.*

Indeed it is our experience that, although for a specific problem instance one solver will be fastest, for most problem classes different solvers do better on different problem instances. Moreover in solving a single problem instance, there are different stages of the problem solving process when different solvers make the fastest progress.

Consequently the most robust algorithm, achieving overall good performance with the best worst-case performance, is a combination of all the solvers, where constraints are posted to *all* the solvers which can handle them. Good performance is achieved because the solvers communicate information with each other. To make this communication possible we require channeling constraints that enable information exported by one solver, expressed in terms of the variables handled by the solver, to be translated into information expressed in terms of a different set of variables that occur in another solver [CLW96].

The behaviour of a channeling constraint between an FD variable and its related boolean variables is not very mysterious. If variable V has domain $1..n$, and there are n boolean variables, B_1 which represents $V = 1$, B_2 which represents $V = 2$ etc., then we have the following propagation behaviour:

- If B_i is instantiated to 1 then propagate $V = i$
- If B_i is instantiated to 0 then propagate $V \neq i$
- If V is instantiated to i then propagate $B_i = 1$
- If the value i is removed from the domain of V, then propagate $B_i = 0$

To complete the channeling constraints we add the constraint $\sum_i B_i = 1$, which reflects that V takes one, and only one, value from its domain.[6]

Channeling constraints support communication between FD, interval, linear, SAT and set solvers. Note that they can also be useful when two design models of the same problem are mapped to the same solver. As a simple example the same n-queens problem can be modelled with a queen for every row, and a queen for every column. Let QR_i denote the position (i.e. column) of the queen in row i, and let QC_m denote the position (i.e. row) of the queen in column m. The channeling constraints are as follows:

---

[6] For mathematical programmers, the B_i comprise an SOS set of type one.

- If QR_i takes the value m, then propagate $QC_m = i$.
- If QC_j takes the value n, then propagate $QR_n = j$
- If m is removed from the domain of QR_i, then propagate $QC_m \neq i$
- If n is removed from the domain of QC_j, then propagate $QR_n \neq j$

For standard models and FD implementations, the combined model, with channeling constraints, has better search behaviour than either of the original or dual model.

4.2   Propagation and Local Cuts

Let us now assume that we have one search routine which posts constraints on the variables at each node of the search tree. These constraints are then communicated to all the solvers through channeling constraints, and each solver then derives some information which it exports to the other solvers. The other solvers then use this information to derive further information which is in turn exported, and this continues until no more new information is produced by any of the solvers.

In this section we shall discuss what information is communicated between the solvers, and how it is used.

Information Communicated to the Linear Solver. When the domain of an FD variable changes, certain boolean variables become instantiated (either to 1 or to 0), and these values can be passed in to the linear solver. (Note that interval solvers use a different form of channeling discussed later in this section.)

The importance of this communication is worth illustrating with a simple example. Consider the constraints $X \in 1..2$, $Y \in 1..2$ and $Z \in 1..2$, and *alldifferent([X, Y, Z])*. In the linear solver this information is represented using six boolean variables $X_1, X_2, Y_1, Y_2, Z_1, Z_2$ and five constraints $\sum_i X_i = 1$, $\sum_i Y_i = 1$, $\sum_i Z_i = 1$, $X_1 + Y_1 + Z_1 \leq 1$, $X_2 + Y_2 + Z_2 \leq 1$. This immediately fails in both the FD and the linear solvers. Suppose now we change the problem, and admit $Z = 3$. The linear solver determines that there is a solution (e.g. $X = 1, Y = 2, Z = 3$), but infers nothing about the values of X, Y or Z.[7] The FD solver immediately propagates the information that $Z = 3$, via the channeling constraint which adds $Z_3 = 1$ to the linear solver.

Sometimes FD variables are represented in the linear solver as continuous variables with the same bounds. In this case only bound updates on the FD variables are passed to the linear solver. An important case of this is branch and bound search, when one of the solvers has discovered a feasible solution with a certain cost. Whilst linear solvers can often find good cost lower bounds - i.e. optimal solutions to relaxed problems that are at least as good as any feasible solution - they often have trouble finding cost *upper* bounds - i.e. feasible but not necessarily optimal solutions to the real problem. The FD solution exports a cost upper bound to the linear solver. This can be used later to prune search when,

---

[7] The linear solver can be persuaded to infer more information only by trying to maximise and minimise each boolean in turn, as mentioned earlier.

after making some new choices, the linear cost lower bound becomes greater than this upper bound. This is the explanation why hybrid algorithms are so effective on the Progressive Party Problem [SBHW95, RWH99]. CP quite quickly finds a solution with cost 13, but cannot prove its optimality. Integer/linear programming easily determines the lower bound is 13, but cannot find a feasible solution.[8]

Not all constraints can immediately be posted to an individual solver. A class of constraints identified by Hooker [HO99] are constraint with the form $FD(\overline{X}) \rightarrow LP(\overline{Y})$. In this simplified form the $\overline{X}$ are discrete variables, and FD is a complex non-linear constraint on these variables. Finding an assignment of the variables that violates this constraint is assumed to be a hard problem for which search and FD propagation may be suitable. Once the variables in $\overline{X}$ have been instantiated in a way that satisfies FD, the linear constraint $LP(\overline{Y})$ is posted to the linear solver. If at any point the linear constraints become inconsistent, then the FD search fails the current node and alternative values are tried for the FD variables. Assuming the class of FD constraints is closed under negation, we can write $-FD(\overline{X})$ for the negation of the constraint $FD(\overline{X})$. This syntax can be used both to express standard FD constraints, by writing $-FD(\overline{X}) \rightarrow 1 < 0$, (i.e. $-FD(\overline{X}) \rightarrow false$), and standard LP constraints by writing $true \rightarrow LP(\overline{Y})$.

Constraints involving non-linear terms (in which two variables are multiplied together), can be initially posted to the interval solver, but as soon as enough variables have become instantiated to make all the terms linear the constraint can be sent to the linear solver as well. In effect the linear constraint is information exported from the interval solver and sent to the linear solver. Naturally this kind of example can be handled more cleverly by creating a version of the original nonlinear constraint where the non-linear terms are replaced by new variables, and posting it to the linear solver. When the constraint becomes linear, this should still be added to the linear solver as it is logically, and mathematically, stronger than the version with the new variables.

For the workforce scheduling problem we can use several hybrid design models linking finite domain and integer/linear constraints. We can, for example, use the FD solver to choose which employees perform which tasks.

We can link these decisions to the linear solver using Hooker's framework, by writing "*If* task i and task j are assigned to the same employee *then* task i must precede task j". We also need a time period of t_{ij} to travel from the location of task i to that of task j. Thus if task i precedes task j we can write $S_j \geq S_i + t_{ij}$, where S_i is the start time of task i, and S_j that of task j.

Now the linking constraint $FD(\overline{X}) \rightarrow LP(\overline{Y})$ has

- $FD(\overline{X}) = assign(task_i, Emp) \wedge assign(task_j, Emp)$, and
- $LP(\overline{Y}) = S_j \geq S_i + t_{ij}$

---

[8] Though Kalvelagen subsequently modelled this problem successfully for a MIP solver.

Information Exported from the Linear Solver

Cost Lower Bounds. For minimisation problems, we have seen that cost upper bounds passed to the linear solver from the FD solver can be very useful. Cost lower bounds returned from the linear solver to the FD solver are equally important.

Hoist scheduling is a cyclic scheduling problem where the aim is to minimise the time for one complete cycle [RW98]. FD can efficiently find feasible schedules of a given cycle, but requires considerable time to determine that if the cycle time is too short, there is no feasible schedule. Linear programming, on the other hand, efficiently returns a good lower bound on the cycle time, so by running the LP solver first an optimal solution can be found quite efficiently by the FD solver. Tighter hybridisation yields even better algorithm efficiency and scalability [RYS02].

Returning to the workforce scheduling example, another approach is to use the full power of integer/linear programming to solve TSPTW (Traveling Salesman Problem with Time Windows) subproblems and return the shortest route covering a set of tasks. In this case we have a linking constraint for each employee, emp_n. Whenever a new task is added to the current set assigned to emp_n the implication constraint sends the whole set to the integer/linear solver which returns an optimal cost (shortest route) for covering all the tasks.

This is an example of a global optimisation constraint (see [FLM04]). Global optimisation constraints are ones for which special solving techniques are available, and from which not only cost lower bounds can be extracted, but also other information.

Reduced Costs. One very useful type of information is reduced costs. As we have seen, a single FD variable can be represented in the linear solver as a set of boolean variables. When computing an optimal solution, the LP solver also computes reduced costs for all those booleans which are 0 at an optimum. The reduced costs provide an estimate of how much worse the cost would become if the boolean was set to 1. This estimate is conservative, in that it may underestimate the impact on cost.

If there is already a cost upper bound, and the reduced cost shows that setting a certain boolean to 1 would push the cost over its upper bound, then we can conclude that this boolean must be 0 - this is termed *variable fixing* in the mathematical programming community. Via the channeling constraints this removes the associated value from the domain of the associated FD variable. Reduced costs therefore enable us to extract FD domain reductions from the linear solver.

Reduced costs can also be used for exporting heuristic information. A useful variable choice heuristic termed *max regret* is to select the variable with the greatest difference between its "preferred" value, and all the other values in its domain. This difference is measured in terms of its estimated impact on the cost, which we can take as the minimum reduced cost of all the other booleans representing values in the domain of this variable.

Relaxed Solution. The most valuable heuristic information exported from the linear solver is the relaxed solution which it uses to compute the cost lower bound. This assignment of values to the variables is either feasible for the whole problem - in which case it is an optimal solution to the whole problem - or it violates some constraints. This information can then focus the search on "fixing" the violated constraints. Most simply this can be achieved by instantiating one of the variables in the violated constraints to another value (perhaps the one with the smallest reduced cost). However a more general approach is to add new constraints that prevent the violation occurring again without removing any feasible solutions.

Fixing Violations by Adding Cuts. To fix the violation we seek a logical combination of linear constraints which exclude the current infeasible assignment, but still admits all the assignments which are feasible for this constraint.

If this is a conjunction of constraints, then we have a *global cut* which can be added to the design model for the problem. An example of this is a subtour elimination constraint, which rules out assignments that are infeasible for the travelling salesman problem.

If it is a disjunction, then different alternative linear constraints can be posted on different branches of a search tree. When, for example, an integer variable is assigned a non-integer value, say 1.5, by the linear solver, then on one branch we post the new bound $X \leq 1$ and on the other branch we post $X \geq 2$.

The challenge is to design these constraints in such a way that the alternation of linear constraint solving and fixing violations is guaranteed, eventually, to terminate.

Fixing Violations by Imposing Penalties. There is another quite different way to handle non-linear constraints within a linear constraint store. Instead of posting a new constraint, modify the cost function so that the next optimal solution of the relaxed problem is more likely to satisfy the constraint. For this purpose we need a way of penalising assignments that violate the constraint, in such a way the penalty reduces as the constraint becomes closer to being satisfied, and becomes 0 (or negative) when it is satisfied.

With each new relaxed solution the penalty is modified again, until solutions are produced where the penalty function makes little, or no, positive or negative contribution, but the constraint is satisfied. In case the original constraints were all linear, we can guarantee that the optimum cost for the best penalty function is indeed the optimum for the original problem. This approach of solving the original problem by relaxing some constraints and adding penalties to the cost function is called "Lagrangian relaxation".

Information Imported and Exported from the One-Way Solver. The one-way solver is an important tool in our problem solving workbench. The solver is used to detect constraint conflicts, and thus help focus search on hard subproblems. A simple example of this is for scheduling problems where the objective is to complete the schedule as early as possible. FD propagation is used to tighten the bounds on task start times. After propagation, each start time variable is assigned a tentative value, which is the smallest value in its domain.

The one-way solver then flags all violated constraints, and this information is used to guide the next search choice.

For many scheduling problems tree search proceeds by posting at each node an ordering constraint on a pair of tasks: i.e. that the start time of the second task is greater than the end time of the first task. We can reify this ordering constraint with a boolean that is set to 1 if and only if the ordering constraint holds between the tasks. We can use these booleans to infer the number of resources required, based on the number of tasks running at the same time.

Interestingly this constraint can be handled by three solvers: the one-way solver, the FD solver and the linear solver. The linear solver relaxes the integrality of the boolean, and simply finds optimum start times. The FD solver propagates the current ordering constraints, setting booleans between other pairs of tasks, and ultimately failing if there are insufficient resources. The one-way solver propagates tentative start time values exported from the linear solver to the booleans. This information reveals resource bottlenecks, so at the next search node an ordering constraint can be imposed on two bottleneck tasks. The ordering constraint is imposed simply by setting a boolean, which constrains the FD, linear and one-way solvers. This approach was used in [EW00].

The one way solver propagates not only tentative values, but also other heuristic, or meta-information. It allows this information to be updated efficiently by updating summary information rather than recomputing from scratch. This efficiency is the key contribution of *invariants* in the Localizer system [VM00].

For example if Sum is the sum of a set of variables V_i, then whenever the tentative value of a variable V_k is updated, from say m to n, the one way solver can efficiently update Sum by changing its tentative value by an amount $n - m$.

Earlier we discussed reduced costs. The "max regret" heuristic can be supported by using the one-way solver to propagate the reduced costs for the booleans associated with all the different values in the domain of a variable. *MaxRegret* for the variable is efficiently maintained as the difference between the lowest and the second lowest reduced cost.

In concluding this section we can illustrate some of the forms of communication discussed above on our solver diagram.

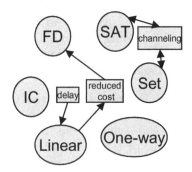

Fig. 2. Explicit communication between solvers

5   Hybrid Algorithms and Decomposed Problems

5.1   Subproblems Handled with Independent Search Routines

Global Cuts and Nogoods. A loose form of hybrid algorithm is to solve two subproblems with their own separate search routines. Each solution to one subproblem is tested against the other by initialising the common variables with the solution to the first subproblem, and then trying to solve the constrained version of the second one. Each time the second subproblem solver fails to find a solution, this is reported back to the first subproblem solver as a nogood. The idea is that future solutions to the first subproblem will never instantiate the common variables in the same way.

This simple idea can be enhanced by returning not just a nogood, but a more generic explanation for the failure in the second subproblem. If the second subproblem is linear, then the impossibility of finding a solution to the second subproblem is witnessed by a linear combination of the constraints $SumExpr \geq SumNumber$ where $SumExpr$ is positive or zero, but $SumNumber$ is negative.

The actual dual values which yield this inconsistency not only show the inconsistency of the given subproblem, but may witness the inconsistency of other assignments to the shared variables. By replacing the values of the shared variables with new values, and combining the linear constraints in the second subproblem using the same dual values we get a new linear constraint called a *Benders Cut* which can be posted to the first subproblem as an additional constraint. Benders Cuts can be used not only to exclude inconsistent subproblems, but also subproblems which cannot participate in an optimal solution to the whole problem [HO03, EW01].

Constraining the Second Subproblem - Column Generation. Curiously there is a form of hybridisation where an optimal solution to the first problem can be used to generate "nogood" constraints on the second problem. This is possible when any solution to the first problem is created using a combination of solutions to the second problem.

This is the case, for example where a number of tasks have to be completed by a number of resources. Each resource can be used to perform a subset of the tasks: finding a set of tasks that can be performed by a single resource (a "line of work") is the subproblem. Each line of work has an associated cost. The master problem is to cover all the tasks by combining a number of lines of work at minimal cost. Given an initial set of lines of work, the master problem is to find the optimal combination. The shadow prices for this solution associate a price with each task. A new line of work can only improve the optimum if its cost is less than the sum of the shadow prices of its set of tasks. Such lines of work are then added to the master problem. They appear as columns in the internal matrix used to represent the master problem in the LP solver. This is the reason for the name *Column Generation*. This is the requirement that is added as a constraint on the subproblem.

This technique applies directly to the workforce scheduling problem. A preliminary solution is computed where each employee performs just one task, for which he/she is qualified. This solution is then improved by seeking sets of tasks that can feasibly be performed by an employee, at an overall cost which is less than the sum of the shadow prices associated with the tasks. The master problem then finds the best combination of employee lines of work to cover all the tasks. This optimal solution yields a modified set of shadow prices which are used to constraint the search for new improving lines of work. This continues until there are no more lines of work that could improve the current optimal solution to the master problem.

5.2   Design Models for Hybrid Algorithms

We briefly list the steps in building design models for hybrid algorithms.

1. Associate constraints with solvers, and initialise variables
2. Add channeling constraints and explicit solver communication
3. Write problem decompositions, and communication between submodels
4. Specify search for each submodel

The specification of the search algorithm is a big part of writing the design model. Search will be covered in section 6 below.

5.3   Applications

Perhaps the first hybrid application using the CP framework was an engineering application involving both continuous and discrete variables [VC88]. The advantages of a commercial linear programming system in conjunction with constraint programming was discovered by the Charme team, working on a transportation problem originally tackled using CP. Based on this experience we used a combination of Cplex and ECLiPSe on some problems for BA [THG+98]. This worked and we began to explore the benefits of the combination on some well known problems [RWH99].

Hybrid techniques proved particularly useful for hybrid scheduling applications [DDLMZ97, JG01, Hoo05, BR03, RW98, EW00]. Hybridisation techniques based on Lagrangian relaxation and column generation were explored, as a way of combining LP and CP [SZSF02, YMdS02, SF03].

Finally these techniques began to be applied in earnest for real applications at Parc Technologies in the area of networking [EW01, CF05, OB04] and elsewhere for sports scheduling [ENT02, Hen01].

6   Hybrid Search

6.1   Context

Search is needed to construct solutions to the *awkward* part of any combinatorial problem. When the constraint solvers have done what inferences they can, and have reached a fixpoint from which no further information can be extracted, then the system resorts to search. Our final solver illustration shows information communicated from the solvers to the search engine.

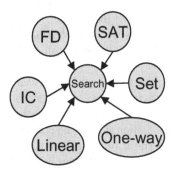

Fig. 3. Solvers and Search

Example kinds of information communicated to the search engine are:

FD	Variable domain size
IC	Variable domain width
Linear	Variable regret
One-way	Variable tentative value
Set	Possible members
SAT	Variable heuristic measure

Because search is guided by heuristics, it is impossible to say a priori that one search algorithm is more appropriate for a certain problem than another. There is always a chance that the non-recommended search algorithm happens to chance on an optimal solution straightaway. Moreover there are almost endless possibilities for search hybridisation and one cannot hope to survey them all. The objective here is simply to paint the big picture, or create a global map, so that different techniques can be located, and their relationships can be understood.

6.2 Overview

We distinguish two main search methods:

- Constructive search, which adds constraints incrementally to partial solutions
- Local search, which seeks to improve a (set of) complete assignments of values to the variables by making small "local" changes

Constructive Search. The simplest form of constructive search is *greedy* search, which constructs a solution incrementally by adding constraints, but never reconsiders any alternatives in the light of new information (such as a dead-end). To achieve completeness it is necessary to explore alternative choices using backtracking [Gas74], or memory-intensive techniques that maintain a whole frontier of nodes yet to be explored [KZTH05].

Another form of constructive search is *labelling*, where the constraint added at each step equates a variable to a specific value. Clearly labelling can also be "greedy".

The general form of constructive search is *tree search*. Under each node of the tree, except the *leaf* nodes, there are several branches, and a different constraint is added at each branch. Each branch ends at another node.

The tree is complete if the disjunction of the constraints added on the branches under a node is entailed by the subproblem at the node. The subproblem at a node is a conjunction of the original problem with all the constraints added on its ancestor branches. The purpose of search is to add constraints down a branch until the solver is guaranteed to detect if the constraints are inconsistent. This property is called *solver completeness*. At each leaf of the search tree, therefore the solver must be complete for the conjunction of problem constraints and constraints imposed down the branch [MSW06].

Optimisation can be achieved using tree search. This could be achieved by finding all solutions and keeping the best. However the *branch and bound* method improves on this by adding, after each new solution is found, a constraint that requires future solutions to be better than the current best one. After adding a new constraint like this, search can continue on the original search tree - since the constrained search tree is a subtree - or it can restart with a new search tree [Pre99].

The branches in a search tree can be explored in different orders. So even for complete search we can distinguish depth-first search, breadth-first, best-first, and others [KZTH05].

A search tree can be explored completely, or incompletely. Indeed greedy search can be seen as a form of incomplete tree search. The are many ways of limiting the search effort, to yield an incomplete tree search. The simplest is to stop after a specified maximum time. Similarly the number of backtracks can be limited, or a limited amount of search *credit* can be allocated to the search algorithm, and the algorithm can share that credit - which represents search effort - in different ways among the subtrees [BBCR97]. For example beam search is defined in Wikipedia as follows: It is a heuristic search algorithm that is an optimization of best-first search. Like best-first search, it uses a heuristic function to estimate the promise of each node it examines. Beam search, however, only unfolds the first m most promising nodes at each depth, where m is a fixed number, the "beam width."

Local Search. For local search we need to associate with each complete assignment a value which we will call its *price*. The simplest form of local search is *Monte Carlo* search, which just tries different complete assignments at random, and just keeps the ones with the highest price. *Hill climbing* introduces the concept of a *neighbour*. Each neighbour of a complete assignment A is nearly the same as A up to a local change. The hill climbing algorithm moves from an assignment A to one of its neighbours B if B has a higher price than A. Hill-climbing stops when the current assignment has no neighbours with a higher price. There are variants of hill climbing where "horizontal" moves are allowed, and variants where at each step the neighbour with the highest price is chosen [SR95].

The simplest kind of local change is to change the value of a single variable. (This can be seen as the local search equivalent of labelling.) For constrained problems special local changes are introduced which maintain the constraint, such as the two- and three- swaps used for the travelling salesman problem. Sometimes complex changes are used which involve several sequential local changes, for example in Lin and Kernighan's TSP algorithm [LK73]. Ultimately, as for example in *variable neighbourhood search* the change may not be local at all [MH97].

The drawback of hill climbing is that it stops on reaching a "local optimum" which may be far worse than the global optimum. Many forms of local search have been introduced which are designed to escape from local optima. Some of these can work with just one complete assignment, such as Simulated Annealing and Tabu search [BK06]. Others work on a whole population of solutions, combining them in various ways to yield (hopefully) better solutions. Examples include genetic algorithms, ant colony optimisation and several other methods which are inspired by nature [BK06].

6.3   Benefits of Search Hybridisation

The benefits of hybrid search are similar to the benefits of hybrid constraint solving:

- Derive more information about feasibility
- Derive more information about optimality
- Derive more heuristic information

In principle tree search is useful for providing information about feasibility, and local search for providing information about optimality. Local search is also an excellent source of heuristic information.

Probably the best illustration of the potency of hybrid search algorithms is to look at the currently best-performing algorithms on well-known benchmarks such as vehicle routing [HG05], and job shop scheduling [EDU04]. The best solutions to complex real-world benchmarks such as the ROADEF problems [ROA07] are also hybrid.

Ideally this section would also include a guide to writing design models for expressing a wide variety of hybrid search techniques. The nearest there is to such a modeling formalism is perhaps the Comet language [VM05], which is more a programming language than a modelling language. The reason is that as yet it has not been possible to design a design modeling formalism for hybrid search which combines precision with a high level of abstraction.

6.4   Hybrid Solvers and Search

In this section, we will start by exploring how hybrid constraint solvers feed information to a non-hybrid constructive and local search.

Constructive Search. Clearly all the active solvers perform inference at each node of the search tree, inferring new information from the extra constraint posted on the previous branch.

In addition to inferring logical information, the solvers can export heuristic information. The FD solver can export domain sizes and the linear solver can export reduced cost information to be used for a max regret variable choice heuristic.

Moreover the linear relaxed solution exported from the linear solver, may be propagated onto other variables by the one-way solver, and the new tentative values can contribute to constraint violations of measurable size. This can serve as another variable choice heuristic.

To this point we have made little mention of *value* choice heuristics. One source of value choice heuristics is the previous solution, in a changed problem which results from modifying the previous one. Another source is a solution to a relaxed (typically polynomial) subproblem, such as the subproblem defined by the linear constraints only. Let us suppose there are different value choice heuristics exported to the solver. When the heuristics agree, this is a pretty powerful indicator that the heuristic value is the best one.

Limited Discrepancy Search [HG95] is an incomplete constructive search method where only a limited number of variables are allowed to take non-preferred values. An idea due to Caseau is only to count the preferences when we are confident about the heuristic [CLPR01]. Using multiple value choice heuristics, we only count discrepancies from the heuristic suggestion *when the different value choice heuristics agree.* This agreement can also be used as a variable choice heuristic: label first those variables whose value choices we are confident of.

Local Search. One important mechanism for escaping from local optima, in a local search algorithm, is to increase the penalty of all constraints violated at this local optimum. This changes the optimisation function until the assignment is no longer a local optimum [VT99]. The global optimum can eventually be found by hill-climbing when the constraint penalties are just right.

We have already encountered the requirement to find the right penalties in an LP framework: Lagrangian relaxation. The penalty optimisation is often performed by a local improvement technique termed *subgradient optimisation.*

We have essentially the same technique within a local search setting [SW98, CLS00]. It would be interesting to characterise the class of problems for which ideal constraint penalties exist, that have a gradient at every point. Clearly LP problems have this property, but are there larger classes?

6.5   Loose Search Hybridisation

Constructive then Local Search. The simplest and most natural form of search hybridisation is to use a constructive search to create an initial solution, and then to use local search to improve it. Local search routines need an initial solution to start with, and the quality of the initial solution can have a very significant impact on the performance of the local search. Constructive search is the only way an initial solution can be constructed, of course! Typically a greedy search is used to construct the initial solution, and constraint propagation plays a very important role in maximising the feasibility of the initial solution. However when the domain of a variable becomes empty instead of failing the greedy search method chooses for that variable a value in conflict and continues with the remaining variables.

For industrial applications where constructive search is used as the heart of the algorithm, because of its suitability for dealing with hard constraints, there is a risk that the final solution suffers from some obvious non-optimality. If any users of the system can see such "mistakes" in the solution constructed by the computer, there is a loss of confidence. To avoid this, for many industrial applications which are handled using constructive search, a local search is added at the end to fix the more obvious non-optimalities.

For example in a workforce scheduling problem, the final solution can be optimised by trying all ways of swapping a single task from one employee to another and accepting any swap that improves the solution.

Local Search then Constructive. This is a rarer combination. The idea is that the local search procedure reaches a "plain" - an area where further improvement is hard to achieve. Indeed statistical analysis of local improvement procedures show a rate of improvement which decreases until new, better, solutions are rarely found.

At this point a change to a complete search procedure is possible. Indeed, by learning which variable values have proven their utility during the local search procedure, the subsequent complete search can be restricted only to admit values with a higher utility, and can converge quickly on better solutions than can be found by local search [Li97].

More generally, constructive branch and bound algorithms typically spend more time searching for an optimal solution than on proving optimality. Often after finding an intermediate best solution, the search "goes down a hole" and takes a long time to find a better solution. After a better solution is found, then the added optimisation constraint enables a number of better and better solutions are found quite quickly, because a large subtree has been pruned by the constraint. Consequently, using local search to quickly elicit a tight optimisation constraint can be very useful for accelerating the constructive branch and bound search.

This combination is typically used where a proof of optimality is required.

6.6   Master-Slave Hybrids

A variety of master/slave search hybrids are applied to a didactic transportation problem in an interesting survey paper [FLL04].

Constructive Search Aided by Local Search. As a general principle, the inference performed at each node in a constructive search may be achieved using search itself. For example when solving problems involving boolean variables, one inference technique is to try instantiating future boolean variables to 1 and then to 0, and for each alternative applying further inference to determine whether either alternative is infeasible. If so, the boolean variable is immediately instantiated to the other value. This kind of inference can exploit arbitrarily complex subsearches.

Local search can be used in the same way to extract heuristic information. One use is as a variable labelling heuristic in satisfiability problems. At each node in the constructive search tree, use local search to extend the current partial solution as far as possible towards a complete solution. The initial solution used by local search can be the local search solution from a previous node. The variable choice heuristic is then to choose a variable in conflict in this local search solution [WS02].

Another use as a value choice heuristic is to extend each value choice to an optimal solution using local search, and choose the value which yields he best local search optimum. Combining this with the above variable choice heuristic results in a different local search optimum being followed at each node in the constructive search tree.

A quite sophisticated example of this hybridisation form is the "local probing" algorithm of [KE02]. The master search algorithm is a constructive search where at each node a linear temporal constraint is added to force apart two tasks at a bottleneck. This is similar to the algorithm mentioned in Section 4.2 above [EW00]. However the slave algorithm is a local algorithm which performs simulated annealing to find a good solution to the temporal subproblem. The resource constraints are handled in the optimisation function of the subproblem. Moreover this is a three-level hybridisation, because the local move used by the simulated annealing algorithm is itself a constructive algorithm. This bottom level algorithm creates a move by first arbitrarily changing the value of one variable. It then finds a feasible solution as close as possible to the previous one, but keeping the new value for the variable. If there is a single feasible solution constructible from the initial value assignment, then the move operator is guaranteed to find it. This algorithm is therefore complete in the sense that it guarantees to find a consistent solution if there is one, but it sacrifices any proof of optimality.

Local Search Aided by Constructive Search. When neighbourhoods are large, or if the search for the best neighbour is non-trivial, then constructive search can be used effectively for finding the next move in a local search algorithm.

One method is to use constraint propagation to focus the search for the best neighbour [PG99]. A more specialised, but frequently occurring, problem decomposition is to have local search find values for certain key variables, and allow constructive search to fill in the remaining ones. Indeed the simplex algorithm is an example of this kind of hybrid. For workforce scheduling we might use local search to allocate tasks to employees and constructive search to create an optimal tour for each employee. In this hybrid form the constructive search is, in effect, calculating the cost function for the local search algorithm.

An example of this kind of decomposition is the very efficient tabu search algorithm for job shop scheduling described in [NS96]. A move is simply the exchange of a couple of tasks on the current solution's critical path. This is extended to a complete solution using constructive search.

The "combine and conquer" algorithm [BB98] is an example of this hybrid where a genetic algorithm is used for local search. The genetic algorithm works not on complete assignments, but on combinations of subdomains, one for each variable. A crossover between two candidates is a mixing and matching of the subdomains. The quality of a candidate is determined by a constructive search which tries to find the best solution possible for the given candidate within a limited time.

6.7   Complex Hybrids of Local and Constructive Search

There has been an explosion of research in this area over the last decade. Papers have been published in the LS research community, the SAT community, the management science community and others. Some interesting collections include *The Knowledge Engineering Review, Vol 16, No. 1* and the *CPAIOR* annual conference.

In this section we will review two main approaches to integrating constructive and local search:

– Interleaving construction and repair
– Local search in the space of partial solutions

Interleaving Construction and Repair. Earlier we discussed how to apply local search to optimise an initial solution produced by a constructive search. However during the construction of the initial solution, the search typically reaches a node where all alternative branches lead to inconsistent subnodes. Instead of completing the construction of a complete infeasible assignment, some authors have proposed repairing the partial solution using local search [JL00].

This is applicable in case labelling used for the constructive search: extending the approach to other forms of constructive search is an open research problem. The local search is initiated with the first constructed infeasible partial assignment, and stops when a feasible partial assignment has been found. Then constructive search is resumed until another infeasible node cannot be avoided. This interleaved search continues until a feasible solution is found.

I have applied this technique in a couple of industrial applications. Each time the neighbourhood explored by the local search has been designed in an application-specific way. An assignment involved in the conflict has been changed, so as to remove the conflict, but this change often causes a further conflict, which has to be fixed in turn. This iterated fixing is similar to *ejection chains* used in vehicle routing problems [RR96].

Caseau noted that this interleaving of construction and local search yields faster optimisation and better solutions than is achieved by constructing a complete (infeasible) assignment first and then applying local search afterwards [CL99]. Again this observation is borne out by my experience in an industrial crew scheduling problem, where previous approaches using simulated annealing produced significantly worse solutions in an order of magnitude longer execution time.

Local Search with Consistent Partial Solutions. There is a long line of research driven by the problem of what to do when a constructive search reaches a node whose branches all lead to inconsistent subnodes.

Weak commitment search [Yok94] stops the constructive search, and records the current partial assignment as tentative values. Constructive search then restarts at the point of inconsistency, minimising conflicts with the tentative values, but where conflict cannot be avoided, assigning new values to them. Each time a dead-end is reached the procedure iterates. Theoretical completeness of the search procedure is achieved by recording *nogoods*. An incomplete variant is to forget the nogoods after a certain time, effectively turning them into a tabu list.

Another related approach - which was introduced at the same conference - is to commit to the first value when restarting, and then try to label variables, as before, in a way consistent with the tentative variables. In this way all the variables eventually become committed, and completeness is achieved by trying alternatives on backtracking [VS94].

More recently, however, researchers have been prepared to sacrifice completeness, but have kept the principle of maintaining consistent partial solutions. In this framework a local move is as follows

1. Extend the current partial solution consistently by instantiating another variable
2. If no consistent extension can be found, uninstantiate a variable in the current partial solution

This approach was used in [Pre02], and termed *decision repair* in [JL00].

This framework can be explored in many directions:

– What variable and value choice heuristics to use when extending consistent partial solutions
– What propagation and inference techniques to use when extending consistent partial solutions
– What explanations or nogoods to record when a partial solution cannot be extended
– What variable to uninstantiate when a partial solution cannot be extended

In principle this framework can be extended to allow inconsistent partial solutions as well. With this tremendous flexibility almost all forms of search can be represented in the framework, from standard labelling to hill climbing [PV04].

7 Summary

Hybrid techniques are exciting and an endless source of interesting research possibilities. Moreover they are enabling us to take great strides in efficiency and scalability for solving complex industrial combinatorial optimisation problems.

Unpredictability is perhaps the greatest practical problem we face when solving large scale combinatorial optimisation problems. When we first tackled the hoist scheduling problem using a hybrid FD/linear solver combination, what really excited me was the robustness of the algorithm for different data sets.

The downside is that it is hard to develop the right hybrid algorithm for the problem at hand. The more tools we have the harder it is to choose the best combination [WS02].

I believe that after an explosion of different algorithms, frameworks such as that of [PV04] will emerge and we will have a clear oversight of the different ways of combining algorithms. Eventually, a very long time in the future, we may even be able to configure them automatically. That is the holy grail for combinatorial optimisation.

References

[AAH95]   Azarmi, N., Abdul-Hameed, W.: Workforce scheduling with constraint logic programming. BT Technology Journal 13(1) (1995)

[AdVD+04]   Ansótegui, C., del Val, A., Dotú, I., Fernàndez, C., Manyà, F.: Modeling choices in quasigroup completion: Sat vs. csp. In: AAAI, pp. 137–142 (2004)

[AW06]   Apt, K., Wallace, M.: Constraint Logic Programming Using ECLiPSe. Cambridge University Press, Cambridge (2006)

[BB98]   Barnier, N., Brisset, P.: Combine and conquer: Genetic algorithm and cp for optimization. In: CP '98: Proceedings of the 4th International Conference on Principles and Practice of Constraint Programming, London, UK, p. 463. Springer, Heidelberg (1998)

[BBCR97]   Beldiceanu, N., Bourreau, E., Chan, P., Rivreau, D.: Partial search strategy in CHIP. In: Proceedings of the 2nd. International Conference on Meta-Heuristics (1997)

[BK06]   Burke, E., Kendall, G. (eds.): Search Methodologies: Introductory Tutorials in Optimization and Decision Support Methodologies. Springer, Heidelberg (2006)

[BR03]   Beck, C., Refalo, P.: A Hybrid Approach to Scheduling with Earliness and Tardiness Costs. Annals of Operations Research 118, 49–71 (2003)

[CF05]   Cronholm, W., Ajili, F.: Hybrid branch-and-price for multicast network design. In: Proceedings of the 2nd International Network Optimization Conference (INOC 2005), pp. 796–802 (2005)

[CL99]   Caseau, Y., Laburthe, F.: Heuristics for large constrained vehicle routing problems. Journal of Heuristics 5(3) (1999)

[CLPR01]   Caseau, Y., Laburthe, F., Pape, C.L., Rottembourg, B.: Combining local and global search in a constraint programming environment. Knowl. Eng. Rev. 16(1), 41–68 (2001)

[CLS00]   Choi, K.M.F., Lee, J.H.M., Stuckey, P.J.: A lagrangian reconstruction of genet. Artif. Intell. 123(1-2), 1–39 (2000)

[CLW96]   Cheng, B.M.W., Lee, J.H.M., Wu, J.C.K.: Speeding up constraint propagation by redundant modeling. In: Principles and Practice of Constraint Programming, pp. 91–103 (1996)

[DDLMZ97]   Darby-Dowman, K., Little, J., Mitra, G., Zaffalon, M.: Constraint logic programming and integer programming approaches and their collaboration in solving an assignment scheduling problem. Constraints 1(3), 245–264 (1997)

[EDU04]   Mehta, S., Demirkol, E., Uzsoy, R.: A computational study of shifting bottleneck procedures for shop scheduling problems. Journal of Heuristics 3(2), 1381–1231 (2004)

[ENT02]   Easton, K., Nemhauser, G., Trick, M.: Solving the Travelling Tournament Problem: A Combined Integer Programming and Constraint Programming Approach. In: Burke, E.K., De Causmaecker, P. (eds.) PATAT 2002. LNCS, vol. 2740, Springer, Heidelberg (2003)

[EW00]   El Sakkout, H., Wallace, M.: Probe backtrack search for minimal perturbation in dynamic scheduling. Constraints 5(4) (2000)

[EW01]   Eremin, A., Wallace, M.: Hybrid benders decomposition algorithms in constraint logic programming. In: Walsh, T. (ed.) CP 2001. LNCS, vol. 2239, pp. 1–15. Springer, Heidelberg (2001)

[FLL04]   Focacci, F., Laburthe, F., Lodi, A.: Local search and constraint programming. In: Constraint and Integer Programming Toward a Unified Methodology. Operations Research/Computer Science Interfaces Series, chapter 9, vol. 27, Springer, Heidelberg (2004)

[FLM04]   Focacci, F., Lodi, A., Milano, M.: Exploiting relaxations in CP. In: Constraint and Integer Programming Toward a Unified Methodology. Operations Research/Computer Science Interfaces Series, chapter 5, vol. 27, Springer, Heidelberg (2004)

[Gas74]   Gaschnig, J.: A constraint satisfaction method for inference making. In: Proc. 12th Annual Allerton Conf. on Circuit System Theory, pp. 866–874, Univ. Illinois (1974)

[Ger01]   Gervet, C.: Large scale combinatorial optimization: A methodological viewpoint. DIMACS Series in Discrete Mathematics and Computers Science 57, 151–175 (2001)

[Hen01]   Henz, M.: Scheduling a major college basketball conference–revisited. Oper. Res. 49(1), 163–168 (2001)

[HG95]   Harvey, W.D., Ginsberg, M.L.: Limited discrepancy search. In: Proceedings of the Fourteenth International Joint Conference on Artificial Intelligence, pp. 607–615 (1995)

[HG05]   Homberger, J., Gehring, H.: A two-phase hybrid metaheuristic for the vehicle routing problem with time windows. Eur. J. Oper. Res. 162, 220–238 (2005)

[HO99]   Hooker, J.N., Osorio, M.A.: Mixed logical / linear programming. Discrete Applied Mathematics 96-97, 395–442 (1999)

[HO03]   Hooker, J.N., Ottosson, G.: Logic-based Benders decomposition. Mathematical Programming 96, 33–60 (2003)

[Hoo05]   Hooker, J.N.: A Hybrid Method for Planning and Scheduling. Constraints 10(4) (2005)

[JG01]   Jain, V., Grossmann, I.: Algorithms for Hybrid MILP/CP Models for a
           Class of Optimization Problems. INFORMS Journal on Computing 13(4),
           258–276 (2001)
[JL00]   Jussien, N., Lhomme, O.: Local search with constraint propagation and
           conflict-based heuristics. In: Proceedings of the Seventh National Con-
           ference on Artificial Intelligence (AAAI 2000), Austin, TX, USA, August
           2000, pp. 169–174 (2000)
[KE02]   Kamarainen, O., El Sakkout, H.: Local probing applied to scheduling. In:
           Principles and Practice of Constraint Programming, pp. 155–171 (2002)
[KZTH05]   Korf, R.E., Zhang, W., Thayer, I., Hohwald, H.: Frontier search. J.
           ACM 52(5), 715–748 (2005)
[Li97]   Li, Y.: Directed Annealing Searc. In: Constraint Satisfaction and Opti-
           misation. PhD thesis, IC-Parc (1997)
[LK73]   Lin, S., Kernighan, B.W.: An effective heuristic algorithm for the travel-
           ing salesman problem. Operations Research 21, 498–516 (1973)
[MH97]   Mladenović, N., Hansen, P.: Variable neighborhood search. Comps. in
           Opns. Res. 24, 1097–1100 (1997)
[MSW06]   Marriott, K., Stuckey, P., Wallace, M.: Constraint logic programming.
           In: Rossi, Beek, v., Walsh (eds.) Handbook of Constraint Programming,
           chapter 12, Elsevier, Amsterdam (2006)
[NS96]   Nowicki, E., Smutnicki, C.: A fast taboo search algorithm for the job
           shop problem. Manage. Sci. 42(6), 797–813 (1996)
[OB04]   Ouaja, W., Richards, E.B.: A hybrid multicommodity routing algorithm
           for traffic engineering. Networks 43(3), 125–140 (2004)
[PG99]   Pesant, G., Gendreau, M.: A constraint programming framework for local
           search methods. Journal of Heuristics 5(3), 255–279 (1999)
[Pre99]   Prestwich, S.: Three clp implementations of branch-and-bound optimiza-
           tion. In: Parallelism and Implementation of Logic and Constraint Logic
           Programming, vol. 2, Nova Science Publishers, Inc. (1999)
[Pre02]   Prestwich, S.: Combining the scalability of local search with the prun-
           ing techniques of systematic search. Annals of Operations Research 115
           (2002)
[PV04]   Pralet, C., Verfaillie, G.: Travelling in the world of local searches in the
           space of partial assignments. In: CPAIOR, pp. 240–255 (2004)
[ROA07]   ROADEF (2007)
[RR96]   Rego, C., Roucairol, C.: A parallel tabu search algorithm using ejection
           chains for vehicle routing. In: Meta-Heuristics: Theory and Applications,
           Kluwer, Dordrecht (1996)
[RW98]   Rodosek, R., Wallace, M.G.: A generic model and hybrid algorithm for
           hoist scheduling problems. In: CP '98: Proceedings of the 4th Interna-
           tional Conference on Principles and Practice of Constraint Programming,
           London, UK, pp. 385–399. Springer, Heidelberg (1998)
[RWH99]   Rodosek, R., Wallace, M.G., Hajian, M.: A new approach to integrating
           mixed integer programming with constraint logic programming. Annals
           of Operations research 86, 63–87 (1999)
[RYS02]   Riera, D., Yorke-Smith, N.: An Improved Hybrid Model for the Generic
           Hoist Scheduling Problem. Annals of Operations Research 115, 173–191
           (2002)

[SBHW95]   Smith, B., Brailsford, S., Hubbard, P., Williams, H.P.: The Progressive
            Party Problem: Integer Linear Programming and Constraint Program-
            ming Compared. In: Montanari, U., Rossi, F. (eds.) CP 1995. LNCS,
            vol. 976, Springer, Heidelberg (1995)
[SF03]   Sellmann, M., Fahle, T.: Constraint programming based lagrangian re-
            laxation for the automatic recording problem. Annals of Operations Re-
            search 118, 17–33 (2003)
[SR95]   Norvig, P., Russell, S.: Informed Search and Exploration. In: Artificial
            Intelligence: A Modern Approach, chapter 4, Prentice-Hall, Englewood
            Cliffs (1995)
[SW98]   Shang, Y., Wah, B.: A discrete lagrangian-based global-search method
            for solving satisfiability problems. Journal of Global Optimization 12(1),
            61–100 (1998)
[SZSF02]   Sellmann, M., Zervoudakis, K., Stamatopoulos, P., Fahle, T.: Crew as-
            signment via constraint programming: Integrating column generation and
            heuristic tree search. Annals of Operations Research 115, 207–226 (2002)
[THG$^+$98] Hajian, M.T., El-Sakkout, H.H., Wallace, M.G., Richards, E.B., Lever,
            J.M.: Towards a closer integration of finite domain propagation and
            simplex-based algorithms. Annals of Operations Research 81, 421–431
            (1998)
[Van98]   Van Hentenryck, P.: A gentle introduction to Numerica. Artificial Intel-
            ligence 103, 209–235 (1998)
[VC88]   Van Hentenryck, P., Carillon, J.-P.: Generality versus specificity: An ex-
            perience with ai and or techniques. In: AAAI, pp. 660–664 (1988)
[VM00]   Van Hentenryck, P., Michel, L.: Localizer: A Modeling Language for Local
            Search. Constraints 5, 41–82 (2000)
[VM05]   Van Hentenryck, P., Michel, L.: Constraint-Based Local Search. MIT
            Press, Cambridge (2005)
[VS94]   Verfaillie, G., Schiex, T.: Solution reuse in dynamic constraint satisfaction
            problems. In: AAAI '94: Proceedings of the twelfth national conference
            on Artificial intelligence, vol. 1, pp. 307–312 (1994)
[VT99]   Voudouris, E., Tsang, E.P.K: Guided Local Search. European Journal of
            Operational Research 113, 469–499 (1999)
[Wil99]   Williams, H.P.: Model Building in Mathematical Programming. Wiley,
            Chichester (1999)
[WS02]   Wallace, M.G., Schimpf, J.: Finding the right hybrid algorithm - a com-
            binatorial meta-problem. Annals of Mathematics and Artificial Intelli-
            gence 34(4), 259–269 (2002)
[YMdS02]   Yunes, T.H., Moura, A.V., de Souza, C.C.: Hybrid Column Generation
            Approaches for Urban Transit Crew Management Problems. Transporta-
            tion Science 39(2), 273–288 (2002)
[Yok94]   Yokoo, M.: Weak-commitment search for solving constraint satisfaction
            problems. In: Proceedings of the 12th National Conference on Artificial
            Intelligence (AAAI-94), Seattle, WA, USA, pp. 313–318 (1994)
[ZMMM01]   Zhang, L., Madigan, C.F., Moskewicz, M.H., Malik, S.: Efficient conflict
            driven learning in a boolean satisfiability solver. In: ICCAD '01: Proceed-
            ings of the 2001 IEEE/ACM international conference on Computer-aided
            design, pp. 279–285. ACM Press, New York (2001)

An Attempt to Dynamically Break Symmetries in the Social Golfers Problem

Francisco Azevedo

CENTRIA, Universidade Nova de Lisboa, Portugal
fa@di.fct.unl.pt

Abstract. A number of different satisfaction and optimisation combinatorial problems have recently been approached with constraint programming over the domain of finite sets, for increased declarativity and efficiency. Such problems where one tries to find sets of values that satisfy some conditions, often present much symmetry on variables and values. In particular, the social golfers problem encompasses many possible symmetries. Allowing symmetric solutions increases search space unnecessarily, thus multiplying solution time. Therefore, ordering constraints have been proposed and incorporated in set solvers. However, such constraints are imposed statically in the global problem model and are unable to detect symmetries that still occur in sub-problems after a partial labelling. In this paper we discuss how to overcome this and present an approach that sequentially labels variables avoiding such symmetries by dynamically disallowing the assignment of other values from the same equivalence class in the golfers problem. Experimental results show that this approach outperforms previous ones, recently achieved by the constraint programming community, namely over sets. Unfortunately, the current method is incomplete and may loose solutions. Nevertheless, results are correct and show that similar techniques can be used efficiently to obtain faster solutions.

1 Introduction

A number of different satisfaction and optimisation combinatorial problems have recently been approached with constraint logic programming (CLP) over the domain of finite sets (CLP(*Sets*)). In fact, such problems are naturally expressed with set constraints, and recent advances on constraint propagation over this particular domain allowed an efficient solving where other techniques, such as integer linear programming, were inadequate or could not obtain good results.

This kind of problems, where one tries to find sets of values that satisfy some conditions, often presents much possible symmetry on variables and values, which can be swapped and still maintain the solution valid. Allowing symmetric solutions increases search space unnecessarily, thus multiplying solution time, possibly by orders of magnitude. The mere use of set variables automatically allows the removal of some symmetry as opposed to other finite domains. Nevertheless, much symmetry is still possible and, therefore, ordering constraints have been proposed and incorporated in set solvers. However, such constraints are imposed statically in the

F. Azevedo et al. (Eds.): CSCLP 2006, LNAI 4651, pp. 33–47, 2007.

global problem model and are unable to detect important symmetries that still occur in sub-problems after a partial labelling. To overcome this source of inefficiency much research has been recently devoted to (static and dynamic) symmetry breaking in general and, in particular, to the social golfers problem [5,7,9,10-12,14,22,24,25,27].

In this paper we present and discuss an example approach that treats the golfers problem globally and sequentially labels variables avoiding such symmetries by disallowing the assignment of considered equivalent values, which are calculated dynamically, in a similar way to GE-trees [26].

We start by briefly discussing set reasoning approaches and evolution of CLP(*Sets*) solvers in the next section. Then, in section 3, we show two example benchmark applications and present results, discussing the applicability of such solvers and how symmetry is handled. In section 4 we show how equivalence classes of values can be applied to break symmetry in the golfers problem, and present compared results that show the effectiveness of this technique. Finally, we conclude in the last section.

2 Set Reasoning

Set constraint solving has been proposed in [23] and formalised in [13] with ECLiPSe (http://eclipse.crosscoreop.com/eclipse/) library *Conjunto*, specifying set domains by intervals whose lower and upper bounds are known sets ordered by set inclusion. Such bounds are denoted as *glb* (for greatest lower bound) and *lub* (least upper bound). The *glb* of a set variable S can be seen as the set of elements that are known to belong to set S, while its *lub* is the set of all elements that can belong to S. Local consistency techniques are then applied using interval reasoning to handle set constraints (e.g. equality, disjointness, containment, together with set operations such as union and intersection). *Conjunto* proved its usefulness in declarativeness and efficiency for NP-complete combinatorial search problems dealing with sets, compared to constraint solving over finite integer domains.

Afterwards, another ECLiPSe set constraints solver library, *Cardinal* [2-4], improved on *Conjunto* by extending propagation on set functions such as cardinality.

Recently, a set solver based on reduced ordered binary decision diagrams (ROBDD) was proposed [15,19], that may compactly represent a set domain considering only the effective set instances that it may assume (instead of a set interval, that can include an exponentially larger than desired number of possible values). ROBDD operations then allow building efficient domain propagators eliminating intermediate variables, being able to reason more globally. This led to significant improvements over other CP approaches on a comparison that their authors performed.

In the next sections we present models and results using such solvers. As usual, these solvers are not complete, which means that generally a search phase must still occur after posting all constraints, in order to find a solution or prove its impossibility. This search process of labelling set domain variables is usually performed differently from labelling integer variables, where each possible value is directly assigned. With set domains, it is possible to achieve less commitment, by trying to successively include or exclude set elements, instead of assigning a definite set at once, which may lead to more failures. So, when instantiating set variable S (with domain $[glb_S, lub_S]$), each value of $lub_S \backslash glb_S$ will be tried for inclusion or exclusion, until S is ground. This can be done with 2 different strategies: 1) trying inclusion first (which we call "element in set" strategy as in [19]); or 2) trying exclusion first ("element not in set").

Often, variables to label are chosen in increasing order of their domain sizes (*first-fail* heuristic). With sets, this procedure may recalculate the next variable to refine, after each element inclusion or exclusion from the current set variable (and respective propagation), according to the new domains.

3 Example Applications

In this section we discuss modelling and solving of CSPlib (www.csplib.org) problems 44 (Steiner triples) and 10 (social golfers) with set reasoning. These typical applications for set constraint solving are the motivating examples to the introduction of the equivalence classes approach to remove symmetries, described in section 4, where we apply it to the social golfers problem.

3.1 Steiner Triples

The ternary Steiner problem of order n consists of finding a set of $n.(n-1)/6$ triples of distinct integer elements in $U = \{1,...,n\}$ such that any two triples have at most one common element. It is a hyper-graph problem coming from combinatorial mathematics [21] where n modulo 6 has to be equal to 1 or 3 [17,20]. One possible solution for $n=7$ is $\{\{1, 2, 3\}, \{1, 4, 5\}, \{1, 6, 7\}, \{2, 4, 6\}, \{2, 5, 7\}, \{3, 4, 7\}, \{3, 5, 6\}\}$. The solution contains $7*(7-1)/6 = 7$ triples. Steiner Triple Systems (STS) are a special case of Balanced Incomplete Block Designs (BIBD). In fact, a STS with n elements is a BIBD(n, $n.(n-1)/6$, $(n-1)/2$, 3, 1).

A CP(*Sets*) approach directly models this problem by representing each triple as a set variable with upper bound U and cardinality 3, and constraining the cardinality of each intersection of a pair of triples to be not greater than 1. Furthermore, since set elements are automatically ordered, much symmetry is naturally eliminated. Nevertheless, symmetry still occurs, since any 2 sets (triples) in a solution can be swapped without affecting it. For that reason, usually CP(*Sets*) models add ordering constraints on the sets variables.

Since these solvers are not complete, a search phase must still be used to find a completely instantiated solution, and the way to do this may be critical on the computation time. Good heuristics for the order in which variables are picked for assignment and for the order in which values are assigned are then essential. For that, one should study the mathematical properties of the problem to find the best strategy. For example, each element must belong to exactly $(n-1)/2$ triples [8]. With this information, a labelling strategy of deciding, for each element in turn, which are the $(n-1)/2$ triples containing it may drastically reduce computation times. Different simple CP approaches (e.g. [6,13,15,19]) often do not go beyond $n=15$, but may go up until $n=31$, in times under 10 minutes, even using such heuristics and global constraints. In addition, best results are obtained with different labelling strategies, and some smaller instances may remain unsolved. All this shows how labelling strategies are crucial in this sort of problems, according to the propagation used. This is so much so, that, in fact, a backtrack-free labelling for this problem is possible, and the solution is available on the internet (perso.wanadoo.fr/colin.barker/lpa/sts.htm and [1])! A labelling method is described based on the mathematical properties of the problem and the fact that Steiner triple systems can be constructed from a Latin square. One can then easily solve instances for n in the order of 1000 and more. We included this example in this paper not only to warn about this fact, which seems to be mostly unknown to the CP community, but also to show the importance of labelling based on the mathematical study of the problem to be solved.

The labelling phase is especially crucial for satisfaction problems where one tries to find a single solution, which is the case that is already solved. Constraints may still be useful in optimisation where one has to explore the whole search space. For that, the *ROBDD* approach of [19] finds all (30) solutions for $n=7$ in 0.1 seconds, and all (840) solutions for $n=9$ in 22.5 seconds. But, in fact, all solutions for each of these 2 instances are basically the same: a solution can be obtained from another simply by renumbering some values. It has been shown that (up to isomorphism) there is a unique solution for these 2 instances [8], which shows the importance of breaking symmetries.

3.2 Social Golfers

As taken from CSPlib: "The coordinator of a local golf club has come to you with the following problem. In her club, there are 32 social golfers, each of whom play golf once a week, and always in groups of 4. She would like you to come up with a schedule of play for these golfers, to last as many weeks as possible, such that no golfer plays in the same group as any other golfer on more than one occasion."

The problem generalizes to that of scheduling g groups of s golfers over w weeks, such that no golfer plays in the same group as any other golfer twice (i.e. maximum socialization is achieved). Thus, the golfers problem, given values g-s-w, can be formally defined as finding a set, *Weeks*, of sets, for the golfers in *Golfers*, such that:

$$\#Weeks = w \wedge \#Golfers = g \times s \wedge \forall_{wk \,\in\, Weeks} \bigcup_{grp \in wk} grp = Golfers . \tag{1}$$

$$\forall_{wk \,\in\, Weeks}(\# wk = g \wedge \forall_{g1, g2 \,\in\, wk, g1 \neq g2}(\# g_1 = s \wedge g_1 \cap g_2 = \varnothing)) . \tag{2}$$

$$\forall_{w1, w2 \,\in\, Weeks, w1 \approx w2} \forall_{g1 \,\in\, w1} \forall_{g2 \,\in\, w2} \#(g_1 \cap g_2) \leq 1 . \tag{3}$$

Actually, the disjointness condition ($g1 \cap g2 = \varnothing$) in (**2**) is redundant in face of the others.

The optimisation problem corresponds to maximising w, given g and s.

A CP(*Sets*) approach can model the g-s-w satisfaction problem with $w*g$ set variables of cardinality s, and if a solution is found try to increment w iteratively, until there are no more solutions.

Again, much symmetry occurs in this problem since any 2 week variables can be swapped, as well as any 2 groups in the same week, and even 2 values in the whole solution (e.g replacing value 1 with 2, and vice-versa, using integers to represent golfers).

In this section we compare 4 CP approaches to this golfers problem: *Cardinal, ic_sets* and *ROBDD*, together with a more elaborate symmetry breaking approach [18] using ILOG integer solver (www.ilog.com/products/solver), which we refer to as *Symm*. *ROBDD* represents the ROBDD-based set domain solver with merged constraints and no intermediate variables; *ic_sets* is the ECLiPSe library for a solver over sets of integers (cooperating with lib(ic), a hybrid integer/real interval arithmetic constraint solver), using a set ordering constraint that had to be implemented using reified constraints. Let us see how each one copes with the mentioned symmetries.

All 4 approaches order groups in each week with appropriate constraints. In particular, the *ROBDD* model uses a global constraint on the whole week, which partitions it from the set of golfers and simultaneously orders the groups. *Cardinal* just uses #< on the minimum of 2 sets, cooperating with the integer solver, thanks to the *Cardinal* facility to constrain set functions, and making use of the knowledge that sets to be ordered (groups) have each 4 elements and are disjoint. Thus, in each week, for 2 groups, G_1 and G_2, G_1 comes before G_2 iff $min(G_1) < min(G_2)$.

Weeks can be ordered by comparisons on the first group. *Cardinal* did not adopt these ordering constraints since, in general, it led to poorer execution times.

Symm ordering constraints apply globally to row and column symmetries on a Boolean matrix model, achieving more propagation.

As for value symmetry, and as mentioned in [18], some values can be fixed a priori. E.g. for problems with 4 groups of 3 golfers, since initially there is no difference between the possible values, the first week can be fixed as:

  { {1,2,3}, {4,5,6}, {7,8,9}, {10,11,12} }.

Then, values of one group (which cannot play together anymore) can be fixed in other weeks as e.g.:

  { {1,2,3}, {4,5,6}, {7,8,9}, {10,11,12} }
  { {1,?,?}, {2,?,?}, {3,?,?}, {?,?,?} }
  { {1,?,?}, {2,?,?}, {3,?,?}, {?,?,?} }

and so on, for the remaining weeks. *Symm* forces these values and refers to such constraints as basic symmetry breaking constraints. *Cardinal* also adopts this initial assignment of values. This makes no difference in finding a first solution since these values would come up naturally as the first values assigned, due to the modelled constraints and the usual ascending value labelling order. But in optimisation problems, or when proving that there is no solution, a lot of backtracking is avoided.

Table 1 shows the results obtained with these 4 approaches for the same instances of [18] and [19] (*Cardinal*: Pentium 4, 2.4 GHz, 480 Mb RAM; *ic_sets* and *ROBDD*: Pentium-M 1.5 GHz, all 3 with a 10 minutes time limit, whereas *Symm* ran on a 1 GHz Pentium III, 256Mb RAM, for a maximum of one hour on each instance). Shaded rows indicate instances with no possible solution. Times on these rows represent the time needed to prove it.

All 3 CLP(*Sets*) approaches use the "element in set" labelling. *Cardinal* and *ROBDD* obtained the best results with a first-fail heuristic, while *ic_sets* and *Symm* used sequential labelling.

Here, *ROBDD* obtains the best results due to its higher propagation having no intermediate variables, which greatly reduces search space. *ROBDD* is the only approach to solve all these instances, although *Symm* has alternative models (based on

Table 1. Time results (in seconds) of different CP approaches for social golfers

g-s-w	Cardinal	ic_sets	ROBDD	Symm
4-3-5	165.63	—	44.4	0.27
4-3-6	94.67	—	29.6	0.30
5-3-6	—	—	2.0	367.00
5-3-7	—	—	28.4	—
5-4-2	0.83	5.3	0.1	0.07
5-4-3	1.89	9.3	0.5	0.12
5-4-4	3.13	10.5	1.3	0.20
5-4-5	28.65	267.3	4.4	0.84
5-5-7	—	—	0.4	19.30
6-3-6	1.20	2.7	2.5	—
6-4-2	1.75	35.5	0.2	0.20
6-4-3	4.62	59.2	2.3	0.34
6-5-4	—	—	171.5	2.01
7-4-2	2.82	70.3	0.6	7.70
7-4-3	6.37	113.6	3.5	0.50
7-4-4	12.46	135.8	21.8	—
7-4-5	17.18	—	54.7	—
8-3-5	1.01	4.1	6.6	—
8-5-2	—	—	3.1	24.70
9-4-4	42.45	22.7	338.4	—

different orderings) [18] that can solve the remaining instances. The presented *Symm* results correspond to its best model (i.e. the one that can solve more instances). Other models can fill the gaps but then leave other instances unsolved, which shows the importance of using different strategies. Thus, in general, results are very dependent on "luck", according to each specific instance. Even with the more consistent *ROBDD*, results fluctuate a bit, and are still far from corresponding to each problem model size. When applying sequential labelling, *ROBDD* does not obtain a solution in 4 instances.

The fact is that much symmetry is still hidden and not taken into account, as we will see in the next section.

4 Breaking Symmetries with Equivalence Classes of Values

Inspired in the examples above, showing the importance of labelling and symmetry according to the properties of the problem, in this section we discuss and present an approach using equivalence classes of values to break symmetries, which we apply to the social golfers problem with a sequential labelling.

4.1 Examples and Definitions

In the example of the previous section for problems with 4 groups of 3 golfers, we have seen that we could start from an initial fixed configuration as:

{ {1,2,3}, {4,5,6}, {7,8,9}, {10,11,12} }
{ {1,?,?}, {2,?,?}, {3,?,?}, {?,?,?} }
{ {1,?,?}, {2,?,?}, {3,?,?}, {?,?,?} }
...

We could fix the first week because weeks are indistinguishable; we could fix values 1, 2 and 3 in the first 3 groups of other weeks because groups within a week are indistinguishable; and we could fix values using integers from 1 to 12 because values are indistinguishable. Furthermore, this was done in a way that the imposed ordering constraints remain valid.

We can now ask: is that all we can do? Could we fix more values? In fact, the first group of the second week can be fixed as {1,4,7} so that we now have:

{ {1,2,3}, {4,5,6}, {7,8,9}, {10,11,12} }
{ {1,4,7}, {2,?,?}, {3,?,?}, {?,?,?} }
{ {1,?,?}, {2,?,?}, {3,?,?}, {?,?,?} }
...

The reason why this is possible is because, after assigning value 1 to the group, values 2 and 3 (former partners of 1) become impossible, and we need to assign two other values that must come from two other different groups of the first week. Since there is no reason to differentiate between the 3 groups {4,5,6}, {7,8,9}, and {10,11,12}, we may pick the first 2, and from these, pick the first element, since, at this point, there is no reason to differentiate values inside these groups. The first possible groups and first (smallest) values are chosen to make it the lexicographically first group among the remaining possible combinations of values. (The fixing of this group was actually already performed in the *Cardinal* model of the previous section.)

From this point on, the situation is not that simple, since if we continue to sequentially assign to each group the first possible value, we would first assign 5 and 8 to group 2, to obtain {2,5,8}, and then {3,6,9} to the third group, which would lead to having only 10,11, and 12 as possible values to the last group, which is impossible. Since there is a solution to the problem, what went wrong?

What happens is that value 5 cannot be forced to be part of the second group, since now not all values in this set domain are indistinguishable. Although it is easy to see that 5 is indistinguishable from its former partner 6, it is not so in relation to value 10, for instance. The reason is that a former partner of 5 has already been assigned in this week (value 4). So, assigning value 5 now results in having assigned already 2 elements of a former group (all of which must be distributed throughout the whole week), while values from group {10,11,12} still have no assignment in this week, which may make it harder to obtain a solution, since 3 different groups must be found for them, and the week is getting "shorter".

But we have seen that 5 is indistinguishable from 6, so, if value 5 is tried and no solution is obtained, there should be no backtracking to value 6, since it will lead to a failure for the same basic reasons. The same happens with values 10, 11 and 12 – they are indistinguishable since they belong to the same group and have not been used anymore. We can also see that value 8 is indistinguishable from 9 and, in fact, these are also indistinguishable from 5 (and consequently from 6), since they share the same characteristics: they are partners of a golfer which has already been assigned in the current week.

There are thus 2 equivalence classes of indistinguishable values at this point: {5,6,8,9} and {10,11,12}. Therefore, only 2 values (at most) need to be tried to check whether a solution exists. When trying each class, the first value can be tried so that eventual ordering constraints remain valid.

What is it then that, in general, makes two values indistinguishable in the social golfers problem? What we have seen so far, takes into account values already assigned to previous weeks and previous groups (if we consider a week to be a sequence of groups), as well as values already assigned to the current group. It thus makes sense to consider classes of values in a sequential labelling context. For that, we model the g-s-w social golfers problem with a sequence of groups S_{ij} as $<S_{1,1}, S_{1,2}, \ldots S_{1,g}, S_{2,1}, S_{2,2}, \ldots S_{2,g}, \ldots S_{w,1}, S_{w,2}, \ldots S_{w,g}>$, where i is the week index and j the group index, and each S_{ij} is a set, also seen as a sequence, of s golfers. We refer to the nth element of S_{ij} as $S_{ij}(n)$.

In a sequential labelling of weeks, groups and group elements, we define $Dom(i,j,n)$ to be a domain of possible values for $S_{ij}(n)$ in position given by i,j and n, as below, according to: values already assigned in the week, $Assigned(i,j,n)$; current group partners, $GPartners(i,j,n)$; and all former partners of these, $Ps(i,j,n)$:

$$Assigned(i, j, n) = \{x : \exists_{j1, n1}((j_1 \times s + n_1 < j \times s + n) \wedge S_{i, j1}(n_1) = x)\} . \tag{4}$$

$$GPartners(i, j, n) = \{x : \exists_{n1 < n} S_{i, j}(n_1) = x\} . \tag{5}$$

$$Partners(v, i) = \{x : \exists_{i1 < i} \exists_{j}(v \in S_{i1, j} \wedge x \in S_{i1, j} \wedge x \neq v)\} . \tag{6}$$

$$Ps(i, j, n) = \bigcup_{v \in GPartners(i,j,n)} Partners(v, i) . \tag{7}$$

$$Dom(i, j, n) = Golfers \setminus (Assigned(i, j, n) \cup Ps(i, j, n)) . \tag{8}$$

where we represent $Golfers$ as the set of integers ranging from 1 to $g*s$, and $Partners(v,i)$ is the set of partners of golfer v in all weeks before i.

We consider two values of a domain of golfers to belong to the same class, when adding an element to a group, when they have the following 2 counts in common:

1- number of partners from those in the domain
2- number of partners from those already assigned to the week

which we may now formally define as follows.

Two values, v_1 and v_2, of $Dom(i,j,n)$, are said to be equivalent, or symmetrical, on assignment to the nth element of group $S_{i,j}$ when both (9) and (10) below apply:

$$\#(Partners(v_1, i) \cap Dom(i, j, n)) = \#(Partners(v_2, i) \cap Dom(i, j, n)) . \tag{9}$$

$$\#(Partners(v_1, i) \cap Assigned(i, j, n)) = \#(Partners(v_2, i) \cap Assigned(i, j, n)) . \tag{10}$$

Condition (9) differentiates class $\{5,6,8,9\}$ from $\{10,11,12\}$ of the example above. We explain the necessity of condition (10) with the following example: consider the 3-2-3 problem instance when trying to label $S_{3,2}(2)$ as:

    { {1,2}, {3,4}, {5,6} }
    { {1,3}, {2,5}, {4,6} }
    { {1,4}, {2,?}, {?,?} }

At this point, only values 3 and 6 are possible, but only value 6 leads to a solution. Condition (9) does not differentiate them since their partners are, respectively, $\{1,4\}$ and $\{4,5\}$, none of these sets sharing a partner with the current domain. The reason why they are not indistinguishable is that one (value 3) has already 2 partners (1 and 4) assigned in this week, whereas the other (value 6) has only one partner already assigned (4), thus leaving more possibilities to the rest of the groups.

The two conditions are both thus necessary conditions of indistinguishability, although their conjunction is not assured to be a sufficient condition, for this particular problem, as we confirmed later.

We thus consider equivalence classes of symmetrical values in a similar fashion as GE-trees [26], but with our own dynamic *ad-hoc* notion of *symmetrical* values, which may include two non-symmetrical (but *similar*) values in the same equivalence class.

4.2 Algorithm

The basic algorithm to solve a *g-s-w* golfers problem using equivalence classes (EC) of values can thus be defined by the 2 procedures in pseudo-code below (with *GolfersEC(g,s,w,S)* as the top goal), where S is the desired solution as a 3-dimensional array with the complete schedule, if successful:

```
Procedure GolfersEC (In: g,s,w, Out: S)
   for i from 1 to w do
      for j from 1 to g do
         S_{i,j}(1) ← first_value(Dom(S_{i,j}(1)))
         for n from 2 to s do
            AssignEC(S,i,j,n,Dom(i,j,n),∅)
         end for
      end for
   end for
end Procedure

Procedure AssignEC (In/Out:S, In: i,j,n,Vals,Tried)
   remove_first v from Vals yielding RestVals
   N1 ← #(Dom(i,j,n) ∩ Partners(v,i))
   N2 ← #(Assigned(i,j,n) ∩ Partners(v,i))
   ChoicePoint
      (<N1,N2> ∉ Tried
       S_{i,j}(n) ← v
      ) or
      AssignEC(S,i,j,n, RestVals, Tried ∪ {<N1,N2>})
   end ChoicePoint
end Procedure
```

The algorithm consists of a sequential labelling, as explained, with possible choice-points in assignments when there are differentiated values. This is performed on procedure *AssignEC*, by verifying on backtracking whether a new value (from the initial domain) is of a class already tested (in which case, it will not be tried again). Hence, equivalence classes need not be computed at each instantiation – only if necessary will a value be checked for indistinguishability, by verifying the inclusion of its class characteristics in a set (*Tried*) of already tried ones. This method corresponds to using nogoods [16].

The first week is thus automatically fixed and ordered with no possible backtracking, since values are indistinguishable throughout the week. Order of groups within weeks is assured by assigning the first possible value of the domain (which shortens along the week groups) with no possible backtracking, when reaching each first position of a group. This breaks symmetry of groups, since they could be permutated without affecting the solution (hence, no solution is lost here).

An assignment fails when the domain is empty. That is when backtracking takes place, until a solution is found. If the search space is exhausted with no solution found, the procedure fails, indicating that the problem is impossible.

This algorithm, as is, does not guarantee that weeks are sorted. That can be done optionally by comparisons on the first group, but the result is not necessarily better. To obtain a first solution, in general it is better not to force a sorting on weeks. On the other hand, to prove impossibility or in optimisation, where a complete search is needed, it is better to force this sorting since it prunes search space. In fact, our implementation also included this sorting, with a constraint $S_{i,1}(2)$ #< $S_{i+1,1}(2)$, for each week i. This can be done on the 2nd element of a first group since it is guaranteed that the first element will always take value 1.

We also extended this algorithm further with two particular look-ahead optimizations for this problem:

1. We considered the (fixed) last group of the first week, whose s elements must be spread through s different groups in subsequent weeks. Therefore, when assigning the last element of a group, we verify whether we have to place one of those (in case there are no more groups available for the yet unassigned elements in the current week). E.g., with 4 groups of 3 golfers, the first week is { {1,2,3}, {4,5,6}, {7,8,9}, {10,11,12} }. If we are to assign $S_{2,2}(3)$ in a 2nd week as { {1,4,7}, {2,5,?}, {?,?,?}, {?,?,?} }, it must be value 10 (rather than 8 or 9), because otherwise values {10,11,12} would not fit in this week.

2. In each group, the s (ordered) elements must come from s different groups of the first week. Since the first week is completely ordered (from 1 to the total number of golfers), on each assignment of a group position in subsequent weeks, the corresponding golfer has some maximum number possible (otherwise there would not be enough room in the group to receive members of different groups). E.g., with 5 groups of 4 golfers, the first week is { {1,2,3,4}, {5,6,7,8}, {9,10,11,12}, {13,14,15,16}, {17,18,19,20} }. If we are to assign $S_{2,1}(2)$ in a 2nd week as { {1,?,?,?}, {?,?,?,?}, {?,?,?,?}, {?,?,?,?}, {?,?,?,?} }, it must be less than 13. Hence, on backtracking we do not need to consider values 13 to 20, because such values in that position could not lead to a complete group.

4.3   Results and Limitations

In Table 2 we present the results in seconds of our method (*EC*) to find a solution (or exhaust search space) to each of the instances of Table 1, with a C++ implementation on the same 2.4 GHz machine. We reproduce the results already shown for the other CP approaches, for comparison.

Table 2. Time results of other CP approaches for social golfers in comparison with our Equivalence Classes method

g-s-w	Cardinal	ic_sets	ROBDD	Symm	EC
4-3-5	165.63	—	44.4	0.27	0.01
4-3-6	94.67	—	29.6	0.30	0.00
5-3-6	—	—	2.0	367.00	0.13
5-3-7	—	—	28.4	—	1.12
5-4-2	0.83	5.3	0.1	0.07	0.00
5-4-3	1.89	9.3	0.5	0.12	0.00
5-4-4	3.13	10.5	1.3	0.20	0.01
5-4-5	28.65	267.3	4.4	0.84	0.01
5-5-7	—	—	0.4	19.30	0.00
6-3-6	1.20	2.7	2.5	—	0.01
6-4-2	1.75	35.5	0.2	0.20	0.00
6-4-3	4.62	59.2	2.3	0.34	0.00
6-5-4	—	—	171.5	2.01	0.01
7-4-2	2.82	70.3	0.6	7.70	0.01
7-4-3	6.37	113.6	3.5	0.50	0.01
7-4-4	12.46	135.8	21.8	—	0.01
7-4-5	17.18	—	54.7	—	0.02
8-3-5	1.01	4.1	6.6	—	0.04
8-5-2	—	—	3.1	24.70	0.00
9-4-4	42.45	22.7	338.4	—	0.01

First of all, the reason why *EC* is instantaneous in exhausting search space of instances *4-3-6* and *5-5-7* is that they are trivially impossible as checked by the simple test $w \le (g*s-1) \,//\, (s-1)$, which we added to our system (where $//$ is the integer division). This test has a simple explanation: a golfer, x, has to play with $(s-1)$ different partners in each week, and there are only $(g*s-1)$ other golfers (as already noted also in CSPlib). Nonetheless, *4-3-5* is not trivially impossible, and *EC* exhausts search space with no solutions in just 1 hundredth of a second.

EC finds a solution for all other instances also almost instantaneously (only the *5-3-7* instance required 1.12 seconds). *EC* is always faster than *ROBDD*, generally outperforming it by 2 or more orders of magnitude.

In addition, since our method performs a sequential labelling with no definite number of variables *a priori*, it handles better the optimisation problem of finding the maximum number of weeks. In fact, weeks are constructed sequentially and each new week achieved is based on the search already performed for the previous weeks. Hence, our approach reports, for given numbers g and s, each new improved solution obtained, starting from the first week until a maximum is reached, also taking into account the pre-computed trivial upper bound. Thus, instead of trying in turn a specific number of weeks, the maximum is constructed incrementally. Using this method, we extended results of Table 2 to the more complete and organised Table 3 where, given g and s, an upper bound is computed and solutions keep coming (in a time limit of 5 minutes) until the maximum number of weeks is achieved (marked with $\sqrt{}$) either by reaching the upper bound or by exhausting the search space.

Table 3. Search for optimality with Equivalence Classes

Groups	Group Size	Weeks upper bound	Weeks achieved	Search Space Exhausted	Time
4	3	5	4	√	0.01
	4	5	5	√	0.00
5	3	7	7	√	1.10
	4	6	5	√	0.10
	5	6	6	√	0.03
6	3	8	7		1.36
	4	7	6		123.01
	5	7	5		6.56
7	3	10	6		0.08
	4	9	6		0.05
	5	8	5		0.11
8	3	11	8		0.08
	4	10	5		0.02
	5	9	6		15.63
9	3	13	9		4.29
	4	11	5		0.51

EC finds and exhausts search space in all instances with 4 and 5 groups (note that problems with $s>g$ are also trivially impossible for $w>1$, and are not considered in this table). For more than 5 groups, no search was complete, but more weeks than the maximum values presented by the other approaches were achieved. There were even 4 more weeks found in the instance with g-s values of 8-5, and 3 more in instances 6-4 and 8-3 (this one in just 8 hundredths of a second). In addition to presenting new results, all instances were improved.

Note that a particular instance can still be solved much faster with a specific labelling heuristic in a CP approach, but that heuristic will hardly be useful to a large set of instances. For example, using a heuristic of assigning each golfer in turn, to a group in every week, we obtained a 9-weeks solution for 8-4 (the original CSPlib problem) in 8 seconds with *Cardinal*. But then, many other instances remain unsolved: e.g. just by incrementing group size to 5, not even a 2 weeks solutions was then found to the 8-5 problem. *EC*, although seeming disappointing in this instance, is much less dependent on particular instances of this problem, showing a big consistency.

Unfortunately, the described method is incomplete. In fact, we later found a counter example to the possibility that conditions (**9**) and (**10**) could be sufficient for indistinguishability of golfers: in instance 6-5, after running for 6 minutes, search space was exhausted, finding no better than 5 weeks, when a 6 weeks solution is known. Thus, when we exhaust search space, no better solution can be found with the current algorithm; we may not conclude that the optimum was reached. Nevertheless, obtained results are correct and even the exhaustion of search space of Table 2 entries correspond to real optima. The described technique can be used to obtain fast results, or may serve as heuristics integrated in a complete tool. We obtain good results due to the restricted search space, and to our look-ahead optimisations. Notice that we did not include instances with $s=2$ or with $g=s$ in Table 2, cases where usually solutions are easier to obtain (although not always). In such cases, *EC* is particularly fast; for example, for 7-7 the optimum 8-weeks solution is found and search exhausted in just 0.12 seconds. This result has just recently been recognised as the best solution on Warwick Harvey's page for the social golfers problem (formerly at http://www.icparc.ic.ac.uk/~wh/golf/, but currently unavailable), although a solution was not available there. Such page reports results from several different mathematical and CP sources, with specialised techniques based on mathematical constructions and local search, for instance. Currently we cannot improve such results, but we consider that promising results were obtained with this research work.

5 Conclusions and Further Research

In this paper we presented an approach using equivalence classes of values to break symmetries on highly symmetric problems such as those where the goal is to find, or somehow optimise, a set (or sets) of undifferentiated values for a set (or sets) of undifferentiated entities/variables. Current CP approaches model such problems with a fixed number of distinct variables, each with its own domain, thus lacking flexibility and losing information on the similarity of variables and values. This leads to allowing backtracking to an unnecessary re-exploration of search spaces that share the same characteristics. We showed that posting ordering constraints is clearly not sufficient to avoid this, still leaving much symmetry behind.

The Equivalence Classes scheme checks whether a new value in the domain is worth trying, by verifying whether a similar value has already been tried in that context. If so, then the value is simply discarded, thus breaking that symmetry and pruning search space. For this we introduced the notion of equivalent values on a dynamic execution. We applied this technique in the particular social golfers problem, where a sequential labelling allowed: *a)* an easy verification of equivalent values; *b)* immediate breaking of group symmetry (avoiding propagation); and *c)* the possibility for a solution to grow dynamically, thus facilitating solving the optimisation problem.

Experimental results in a C++ implementation showed improvements of orders of magnitude over other CP approaches with efficient propagators (such as an ROBDD-based one), namely over sets, and with global ordering constraints specially implemented to break symmetries [18]. Nevertheless, further comparison with other dynamic symmetry breaking techniques should yet be performed.

Although the indistinguishable conditions presented for this particular problem are incomplete, results are all valid and extended an already large set of instances to improved solutions in a matter of seconds. We consider that the notion of equivalent values and corresponding classes has been sufficiently illustrated with this example, showing its importance for future research. It also shows that it may be worth to partly relax symmetrical conditions in order to obtain larger equivalence classes, thus reducing search space, even if becoming incomplete. Such relaxation may also allow a faster dynamic computation of symmetry and should be further explored in general problems.

In future, for this particular problem, we intend to compare our method with others that search for all solutions, in order to better assess our search space reduction.

We believe this technique can be further improved since it can possibly benefit from other look-ahead techniques and be integrated with another CP approach using global constraints.

Future research should try to find indistinguishable conditions automatically from problem specification. For that, modelled problem variables and values that are really undifferentiated, should not be given names or numbers, *a priori* (as usually done in CP and other programming approaches). More efficiency may be achieved if allowing more declarative models, nearer natural language, to extract only the necessary information, correctly interpreting things like "a set of *n* golfers for a set of weeks" (capturing the implicit symmetries and also allowing a dynamic number of variables).

References

1. Anderson, I., Honkala, I.: A Short Course in Combinatorial Designs (1997), http://www.utu.fi/h onkala/cover.html
2. Azevedo, F.: Constraint Solving over Multi-valued Logics - Application to Digital Circuits. In: Frontiers of Artificial Intelligence and Applications, vol. 91, IOS Press, Amsterdam (2003)
3. Azevedo, F.: Cardinal: A Finite Sets Constraint Solver. Constraints journal, vol. 12(1), pp. 93–129, KAP (2007)
4. Azevedo, F., Barahona, P.: Modelling Digital Circuits Problems with Set Constraints. In: Palamidessi, C., Moniz Pereira, L., Lloyd, J.W., Dahl, V., Furbach, U., Kerber, M., Lau, K.-K., Sagiv, Y., Stuckey, P.J. (eds.) CL 2000. LNCS (LNAI), vol. 1861, pp. 414–428. Springer, Heidelberg (2000)
5. Barnier, N., Brisset, P.: Solving the Kirkman's Schoolgirl Problem in a Few Seconds. In: Van Hentenryck, P. (ed.) CP 2002. LNCS, vol. 2470, pp. 477–491. Springer, Heidelberg (2002)
6. Beldiceanu, N.: An Example of Introduction of Global Constraints in CHIP: Application to Block Theory Problems. Technical Report TR-LP-49. ECRC, Munich, Germany (1990)
7. Choueiry, Noubir.: On the Computation of Local Interchangeability in Discrete Constraint Satisfaction Problems. In: Proc. AAAI'1998 (1998)
8. Colbourn, C.J., Dinitz, J.H. (eds.): Steiner Triple Systems. CRC Handbook of Combinatorial Designs. vol. 70, pp. 14–15, CRC Press, Boca Raton, FL (1996)
9. Fahle, T., Shamberger, S., Sellmann, M.: Symmetry Breaking. In: Walsh, T. (ed.) CP 2001. LNCS, vol. 2239, pp. 93–107. Springer, Heidelberg (2001)
10. Focacci, F., Milano, M.: Global Cut Framework for Removing Symmetries. In: Walsh, T. (ed.) CP 2001. LNCS, vol. 2239, pp. 75–92. Springer, Heidelberg (2001)
11. Gent, I.P., Harvey, W., Kelsey, T.: Groups and Constraints: Symmetry Breaking During Search. In: Van Hentenryck, P. (ed.) CP 2002. LNCS, vol. 2470, pp. 415–430. Springer, Heidelberg (2002)
12. Gent, I.P., Smith, B.M.: Symmetry Breaking During Search in Constraint Programming. In: Proc. ECAI'2000 (2000)
13. Gervet, C.: Interval Propagation to Reason about Sets: Definition and Implementation of a Practical Language. In: Freuder, E.C. (ed.): Constraints journal, vol. 1(3), pp. 191–244, Kluwer Academic Publishers (1997)
14. Harvey, W.: Symmetry Breaking and the Social Golfer Problem. In: Walsh, T. (ed.) CP 2001. LNCS, vol. 2239, Springer, Heidelberg (2001)

15. Hawkins, P., Lagoon, V., Stuckey, P.J.: Set Bounds and (Split) Set Domain Propagation Using ROBDDs. In: Webb, G.I., Yu, X. (eds.) AI 2004. LNCS (LNAI), vol. 3339, pp. 706–717. Springer, Heidelberg (2004)
16. Van Hentenryck, P.: A Logic Language for Combinatorial Optimization. In: Annals of Operations Research (1989)
17. Kirkman, T.P.: On a problem in combinatorics. Cambridge and Dublin Math. Journal 191–204 (1847)
18. Kiziltan, Z.: Symmetry Breaking Ordering Constraints. PhD thesis, Uppsala University (2004)
19. Lagoon, V., Stuckey, P.J.: Set Domain Propagation Using ROBDDs. In: Wallace, M. (ed.) CP 2004. LNCS, vol. 3258, pp. 347–361. Springer, Heidelberg (2004)
20. Lindner, C.C., Rosa, A.: Topics on Steiner Systems. Annals of Discrete Mathematics, North Holland, vol. 7 (1980)
21. Lueneburg, H.: Tools and Fundamental Constructions of Combinatorial Mathematics, Wissenschaftverlag (1989)
22. Meseguer, P., Torras, C.: Exploiting Symmetries Within Constraint Satisfaction Search. Art.Intell. 129, 133–163 (1999)
23. Puget, J.-F.: PECOS: A High Level Constraint programming Language. In: Proc. Spicis 92, Singapore (1992)
24. Puget, J.-F.: Symmetry Breaking Revisited. In: Van Hentenryck, P. (ed.) CP 2002. LNCS, vol. 2470, pp. 446–461. Springer, Heidelberg (2002)
25. Puget, J.-F.: Symmetry Breaking Using Stabilizers. In: Rossi, F. (ed.) CP 2003. LNCS, vol. 2833, Springer, Heidelberg (2003)
26. Roney-Dougal, C.M., Gent, I.P., Kelsey, T., Linton, S.A.: Tractable symmetry breaking using restricted search trees. In: Proceedings of ECAI-04 (2004)
27. Sellmann, M., Harvey, W.: Heuristic Constraint Propagation. In: Proceedings of CPAIOR'02 workshop, pp. 191–204 (2002)

A Constraint Model for State Transitions
in Disjunctive Resources

Roman Barták[1] and Ondřej Čepek[1,2]

[1] Charles University in Prague, Faculty of Mathematics and Physics
Malostranské nám. 2/25, 118 00 Praha 1, Czech Republic
[2] Institute of Finance and Administration
Estonská 500, 101 00 Praha 10, Czech Republic
{roman.bartak,ondrej.cepek}@mff.cuni.cz

Abstract. Traditional resources in scheduling are simple machines where the limited capacity is the main restriction. However, in practice there frequently appear resources with more complex behaviour that is described using state transition diagrams. This paper presents new filtering rules for constraints modelling the state transition diagrams. These rules are based on the idea of extending traditional precedence graphs by direct precedence relations. The proposed model also assumes optional activities and it can be used as an open model accepting new activities during the solving process.

Keywords: constraint, domain filtering, disjunctive resource, state transition.

1 Introduction

Temporal networks play an important role in planning but they are not used as frequently in scheduling where resource restrictions traditionally play a stronger role. This is reflected in scheduling global constraints, where techniques such as edge-finding or not-first/not-last combine restrictions on time windows with a limited capacity of the resource [1]. Recently, a new category of propagation techniques combining information about relative position of activities with capacity of resources appeared [4]. Also techniques combining information about precedence relations and time windows have been proposed [8]. We believe that integration of temporal networks with reasoning on resources [9,10] will play even more important role as planning and scheduling technologies are becoming closer.

In this paper we propose an extension of precedence graphs by direct precedence relations (A can directly precede B if no activity must be allocated between A and B). This extension is motivated by modelling complex behaviour of resources that is described as a state transition diagram. Such a diagram is in fact a generalisation of set-up times that play an important role in current real-life scheduling problems. As factories are transforming from mass production to more customised production, multi-purpose and hence more complicated machines are used and better handling of setups is becoming important. Space applications are another example where a more complex behaviour of resources is typical. Also over-subscribed problems are more

F. Azevedo et al. (Eds.): CSCLP 2006, LNAI 4651, pp. 48–62, 2007.

frequent nowadays so the scheduling systems should be able to handle optional activities, for example, to decide about rejection of activity that cannot be scheduled feasibly together with other activities. Note also, that optional activities are useful for modelling alternative resources (an optional activity is used for each alternative resource) as well as alternative processes to accomplish a job (each process may consist of one of several different sets of activities). Scheduling systems should also be able to add activities to satisfy the transition scheme, for example to insert a setup activity if necessary.

To summarise the contributions of this paper, we propose two extensions of ordinary precedence graphs: adding direct precedence relations and using optional activities. For such a graph which we call a *double precedence graph* we design incremental filtering rules that keep a transitive closure of the graph, deduce new precedences, and decide (in)validity of activities. Moreover, in contrast to traditional global constraints used in scheduling the proposed model is open, that is, it allows adding new activities to the precedence graph during the solution process.

2 Motivation and Problem Description

In this paper we address the problem of modelling a *disjunctive resource* where activities must be allocated in such a way that they do not overlap in time. We assume that there are precedence constraints between the activities. The precedence constraint A « B specifies that activity A must be before activity B in the schedule. In a simplified form, we can assume that each activity X is assigned a sequence number P_X in the solution and the precedence A « B is equivalent to constraint $P_A < P_B$. Each activity is annotated by a resource state requested for processing the activity and there is a state transition diagram describing transitions between the states. *State transition diagram* is a directed graph where nodes describe the states and arcs describe allowed transitions between the states (Figure 1). The state transition diagram restricts sequencing of activities in the following way: activity A can be scheduled directly before activity B ($P_A + 1 = P_B$) only if there is an arc from the state of A to the state of B in the state transition diagram. To model over-subscribed problems and alternative resources/processes, we introduce *optional activities*. An optional activity has one of the following three statuses. If the activity is not yet known to be or not to be included then it is called *undecided*. If the activity is allocated to the resource then it is called *valid*. If the activity is known not to be allocated to the resource then it is called *invalid*. Regular activities correspond to valid activities. The scheduling task is to decide about (in)validity of the undecided activities in such a way that the valid activities form a sequence satisfying the precedence constraints and restrictions imposed by the state transition diagram.

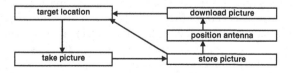

Fig. 1. Example of a state transition diagram

In real-life problems there are usually also time windows restricting position of activity in time. In such a case, it is known that deciding about an existence of a feasible schedule is NP-hard in the strong sense [7] so there is a little hope even for a pseudo-polynomial solving algorithm. Hence using propagation rules and constraint satisfaction techniques is justified here. The paper [2] shows how filtering of time windows can be combined with the precedence graph so in this paper we focus merely on handling (direct) precedence relations. Our goal is to propose filtering rules that remove inconsistencies from the double precedence graph. Namely, the rules deduce new precedence relations and decide whether an activity cannot or must be valid. In fact, the filtering rules implement a global constraint describing allowed sequences of activities.

3 Related Works

Disjunctive temporal networks [11] can model disjunctive resources. However, DTNs use more general disjunctions than necessary and hence they achieve weaker pruning. Moreover, a qualitative approach to time seems more appropriate to describe the problem of activity sequencing. Though we assume durative activities, Interval Algebra is superfluous because disjunctive resources discard most of the interval relations (like starts, during, overlaps etc.). From Point Algebra we need only 'before' and 'after' relations and there is no support for direct precedences there. The work by Laborie [9] studies a combination of resource and temporal reasoning but no algorithm is presented (and a different type of resources is assumed). Probably the closest approach to our problem is presented in paper [6] where alternative resources correspond to paths in the global precedence graph. However, this approach is proposed merely for cost-based filtering (optimization of makespan or setup times) and it assumes all the activities to be present in the global precedence graph. The paper [3] presents an idea of a precedence graph with optional activities. The authors use a so called PEX value to describe a probability of the existence of the activity and their approach is based on updating this value. Instead of that we use a Boolean variable to describe the presence of an activity and we focus more on precedence and direct precedence relations. To summarise the above discussion, none of the existing approaches to temporal and resource reasoning covers fully state transition diagrams and optional activities.

4 Double Precedence Graphs

The precedence relations define a *precedence graph* that is an acyclic directed graph where nodes correspond to activities and there is an arc from A to B if A « B. If access to all predecessors and successors of a given activity is frequently requested, such as in [4,8], then it is more efficient to keep a transitive closure of the graph where this information is available in time O(1) rather than to look for predecessors/successors on demand. Moreover, keeping transitive closure simplifies detection of cycles. We propose the following definition of transitive closure of the precedence graph with optional activities.

Definition 1: We say that a precedence graph G with optional activities is *transitively closed* if for any two arcs A to B and B to C such that B is a valid activity and A and C are either valid or undecided activities there is also an arc A to C in G.

It is easy to prove that if there is a path from A to B such that A and B are either valid or undecided and all inner nodes in the path are valid then there is also an arc from A to B in a transitively closed graph (by induction on the path length). Hence, if no optional activity is used (all activities are valid) then Definition 1 corresponds to a standard definition of the transitive closure.

To model restrictions imposed by the state transition diagram we propose to extend the precedence graph by direct precedence relations between the activities.

Definition 2: We say that A can *directly precede* B if both A and B are either valid or undecided activities, B is not before A ($\neg$ B « A), the transition from A to B is allowed by the state transition diagram, and there is no valid activity C such that A « C and C « B (the relation « is from the transitive closure of the precedence graph with optional activities).

The relation of direct precedence introduces a new type of arc, say «$_d$, in the precedence graph and hence we are speaking about the *double precedence graph*. There is one significant difference between the arcs of type « and the arcs of type «$_d$. While the arcs « are added into the graph as problem solving proceeds, the arcs «$_d$ are typically removed from the graph (note that «$_d$ means "can be directly before", while « means "must be before"). When all valid activities are linearly ordered, there is exactly one arc of type «$_d$ going into each valid activity (with the exception of the very first activity in the schedule) and one arc of type «$_d$ going from each valid activity (with the exception of the very last activity in the schedule).

4.1 Constraint Model

We propose to realise reasoning on precedence relations using constraint satisfaction technology. This allows integration of our model with other constraint reasoning techniques [2]. This integration requires the model to provide full information about precedence relations to all other constraints. We index each activity by a unique number from the set 1, ..., n, where n is the number of activities. For each activity we use a 0/1 variable Valid indicating whether the activity is valid (1) or invalid (0). If the activity is undecided – not yet known to be valid or invalid – then the domain of Valid is {0,1}. The precedence graph is encoded in two sets attached to each activity. CanBeBefore(A) is a set of indices of activities that can be before activity A. CanBeAfter(A) is a set of indices of activities that can be after activity A. For simplicity reasons we will write A instead of the index of A. To simplify description of the propagation rules we define for every activity A the following derived sets:

$$\text{MustBeAfter(A)} = \text{CanBeAfter(A)} \setminus \text{CanBeBefore(A)}$$
$$\text{MustBeBefore(A)} = \text{CanBeBefore(A)} \setminus \text{CanBeAfter(A)}$$
$$\text{Unknown(A)} = \text{CanBeBefore(A)} \cap \text{CanBeAfter(A)}.$$

MustBeAfter(A) and MustBeBefore(A) are sets of those activities that must be after and before the given activity A respectively. Unknown(A) is a set of activities that are not yet known to be before or after activity A (Figure 2).

To model direct precedence relations and hence a double precedence graph, we add two sets to each activity: CanBeRightBefore and CanBeRightAfter containing indexes of activities that can be directly before and directly after a given activity. Naturally, the following relation holds CanBeRightBefore(A) ⊆ CanBeBefore(A) at any time and similarly CanBeRightAfter(A) ⊆ CanBeAfter(A).

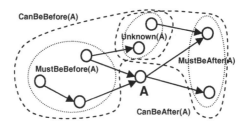

Fig. 2. Representation of the precedence graph

Note on Representation. The main reason for using sets to model the precedence graph is their possible representation as domains of variables in constraint satisfaction packages. Recall that domains of variables can only shrink as problem solving proceeds. The sets in our model are also shrinking as new arcs « are added to the precedence graph. Hence a special data structure is not necessary to describe the precedence graph in constraint satisfaction packages. Moreover, these packages usually provide tools to manipulate the domains, for example membership and deletion operations. In the subsequent complexity analysis, we will assume that these operations require time O(1), which can be realised for example by using a bitmap representation of the sets. Note finally, that empty domain implies inconsistency that may be a problem for the very first and very last activity which has no predecessors and successors respectively. To solve the problem we can simply leave activity A in both sets CanBeAfter(A) and CanBeBefore(A). Then no domain of CanBeBefore and CanBeAfter will ever be empty but we can detect inconsistency via the empty domain of Valid variables.

4.2 Constraint Model

The goal of propagation rules is to remove inconsistent elements (activities) from the above described sets – this is called domain filtering in constraint satisfaction. In the first stage, we will focus on making a transitive closure of the precedence graph according to Definition 1. Note that the transitive closure of the precedence graph also simplifies detection of inconsistency of the graph. The precedence graph is inconsistent if there is a cycle of valid activities. In a transitively closed graph, each such cycle can be detected by finding two valid activities such that A « B and B « A. Our propagation rules prevent cycles by making invalid the last undecided activity in each cycle. This propagation is realised by using an exclusion constraint. When a cycle A « B and B « A is detected, the following exclusion constraint can be posted:

$$Valid(A) = 0 \vee Valid(B) = 0.$$

This constraint ensures that each cycle is broken by making at least one activity in the cycle invalid. Instead of posting the constraint directly to the constraint solver, we propose keeping the set Ex of exclusions. The above exclusion constraint is modelled as a set {A,B} ∈ Ex. Now, the propagation of exclusions is realised explicitly – if activity A becomes valid then all activities C such that {A,C} ∈ Ex are made invalid.

We initiate the precedence graph in the following way. First, the variables Valid(A), CanBeBefore(A), CanBeRightBefore(A), CanBeAfter(A), and CanBeRightAfter(A) with their domains are created for every activity A. Namely, the domain for Valid variable is {0,1} and the domain for other variables is {1,...,n}, where n is the number of activities. Then the known precedence relations in the form A « B are added by removing B from the sets CanBeBefore(A) and CanBeRightBefore(A), and removing A from the sets CanBeAfter(B) and CanBeRightAfter(B). Note, that because all activities are still undecided at this stage, domain change is not propagated to other variables. Finally, the Valid(A) variable for every valid activity A is set to 1 (and similarly Valid variables of invalid activities are set to 0). By instantiating the Valid(A) variable, the propagation rule /1/ is invoked. "Valid(A) is instantiated" is its trigger. The part after → is a propagator describing pruning of domains. "exit" means that the constraint represented by the propagation rule is entailed so the propagator is not further invoked (its invocation does not cause further domain pruning). We will use the same notation in all rules. The propagation rule /1/ realises the above described exclusion constraints as well as adding new arcs according to Definition 1.

```
Valid(A) is instantiated →                                        /1/
  if Valid(A) = 0 then
      Ex := Ex \ {{A,X} | X is an activity}
      for each B do          // disconnect A from B
         CanBeBefore(B)  ← CanBeBefore(B) \ {A}
         CanBeAfter(B)   ← CanBeAfter(B) \ {A}
         CanBeRightBefore(B) ← CanBeRightBefore(B) \ {A}
         CanBeRightAfter(B)  ← CanBeRightAfter(B) \ {A}
  else  // Valid(A)=1
      for each C s.t. {A,C}∈Ex do Valid(C) ← 0
      for each B∈MustBeBefore(A) s.t. Valid(B)≠0 do
         for each C∈MustBeAfter(A) s.t. Valid(C)≠0 do
            CanBeRightAfter(B)  ← CanBeRightAfter(B) \ {C}
            CanBeRightBefore(C) ← CanBeRightBefore(C) \ {B}
            if C∉MustBeAfter(B) then //add arc from B to C
               CanBeAfter(C)  ← CanBeAfter(C) \ {B}
               CanBeBefore(B) ← CanBeBefore(B) \ {C}
               CanBeRightAfter(C)  ← CanBeRightAfter(C) \ {B}
               CanBeRightBefore(B) ← CanBeRightBefore(B) \ {C}
               if C∉CanBeAfter(B) then    // break the cycle
                  if Valid(B)=1 then Valid(C) ← 0
                  // Valid(C)=1 leads to fail
                  else if Valid(C)=1 then Valid(B) ← 0
                     else Ex ← Ex ∪ {{B,C}}
  exit
```

Note that rule /1/ maintains symmetry of sets modelling the double precedence graph for all valid and undecided activities because the domains are pruned symmetrically in pairs. We shall show now, that if the entire precedence graph is known in advance (no arcs are added during the solving procedure), then rule /1/ is sufficient for keeping the transitive closure according to Definition 1.

Proposition 1: Let A_0, A_1, ... , A_m be a path in the precedence graph such that Valid(A_j)=1 for all $1 \le j \le m$-1 and Valid(A_0)≠0 and Valid(A_m)≠0 (that is, the endpoints of the path are not invalid and all inner points of the path are valid). Then $A_0 \ll A_m$, that is, $A_0 \notin$ CanBeAfter(A_m) and $A_m \notin$ CanBeBefore(A_0).

Proof: We shall proceed by induction on m. The base case m=1 is trivially true after initialisation (we assume that for every arc (X,Y) in the precedence graph X is removed from CanBeBefore(Y) and Y is removed from CanBeAfter(X) in the initialisation phase). For the induction step let us assume that the statement of the lemma holds for all paths (satisfying the assumptions of the lemma) of length at most m-1. Let $1 \le j \le m$-1 be an index such that Valid(i_j)←1 was set last among all inner points i_1, ... , i_{m-1} on the path. By the induction hypothesis we get

- $i_0 \notin$ CanBeAfter(i_j) and $i_j \notin$ CanBeBefore(i_0) using the path i_0, ... , i_j
- $i_j \notin$ CanBeAfter(i_m) and $i_m \notin$ CanBeBefore(i_j) using the path i_j, ... , i_m

We shall distinguish two cases. If $i_m \in$ MustBeAfter(i_0) (and thus by symmetry also $i_0 \in$ MustBeBefore(i_m)) then by definition $i_m \notin$ CanBeBefore(i_0) and $i_0 \notin$ CanBeAfter(i_m) and so the claim is true trivially. Thus let us in the remainder of the proof assume that $i_m \notin$ MustBeAfter(i_0).

Now let us show that $i_0 \in$ CanBeBefore(i_j) must hold, which in turn (together with $i_0 \notin$ CanBeAfter(i_j)) implies $i_0 \in$ MustBeBefore(i_j). Let us assume by contradiction that $i_0 \notin$ CanBeBefore(i_j). However, at the time when both $i_0 \notin$ CanBeAfter(i_j) and $i_0 \notin$ CanBeBefore(i_j) became true, that is, when the second of these conditions was made satisfied by rule /1/, rule /1/ must have posted the constraint (Valid(i_0)=0 ∨ Valid(i_j)=0) which contradicts the assumptions of the lemma. By a symmetric argument we can prove that $i_m \in$ MustBeAfter(i_j). Thus when rule /1/ is triggered by setting Valid(i_j)←1 both $i_0 \in$ MustBeBefore(i_j) and $i_m \in$ MustBeAfter(i_j) hold (and $i_m \notin$ MustBeAfter(i_0) is assumed), and therefore rule /1/ removes i_m from the set CanBeBefore(i_0) as well as i_0 from the set CanBeAfter(i_m), which finishes the proof.

□

Proposition 2: The worst-case time complexity of the propagation rule /1/ (instantiation of the Valid variable) including all possible recursive calls is $O(n^2)$, where n is the number of activities.

Proof: If activity A is made invalid then all exclusion pairs that include A are removed from set Ex which could be done in time $O(n)$, if the set is properly implemented (for example as a symmetric $n \times n$ matrix). Moreover, activity A is removed from the sets CanBeBefore, CanBeAfter, CanBeRightBefore, and CanBeRightAfter of all other activities which takes the total time $O(n)$.

If activity A becomes valid then some activities are made invalid and some new arcs may be added to the graph. At most n activities can be invalidated which takes a total time $O(n^2)$. The maximal number of added arcs is $\Theta(n^2)$. It may also happen that some other activities (at most $O(n)$) become invalid to break cycles. However, we already know that the time complexity of making an activity invalid is $O(n)$. Together, the worst-case time complexity to make an activity valid is $O(n^2)$.   □

In some situations arcs may be added to the double precedence graph during the solving procedure, either by the user, by the scheduler/planner, or by other filtering algorithms [2]. The following rule /2/ updates the double precedence graph to keep transitive closure when an arc is added to the double precedence graph. If a new arc A«B is added then we first check whether the arc is not already present in the graph. If it is a new arc then the corresponding sets are updated and a possible cycle is detected (we use the same reasoning as in rule /1/). Finally, if any end point of the arcs is valid, then necessary arcs are added to update the transitive closure according to Definition 1. In such a case, some direct precedence relations are removed according to Definition 2. Note that the propagators for new arcs are evoked after the propagator of the current rule finishes.

```
A«B is added →                                              /2/
    if A∈MustBeBefore(B) then exit        // the arc is already present
    CanBeAfter(B) ← CanBeAfter(B) \ {A}
    CanBeBefore(A) ← CanBeBefore(A) \ {B}
    CanBeRightAfter(B) ← CanBeRightAfter(B) \ {A}
    CanBeRightBefore(A) ← CanBeRightBefore(A) \ {B}
    if A∉CanBeBefore(B) then      // break the cycle
        if Valid(A)=1 then Valid(B)          // Valid(B)=1 leads to fail
        else if Valid(B)=1 then Valid(A) ← 0
                // Valid(A)=1 leads to fail
                else Ex ← Ex ∪ {{A,B}}
    else            // transitive closure
        if Valid(A)=1 then
            for each C∈MustBeBefore(A) s.t. Valid(C)≠0 do
                CanBeRightAfter(C) ← CanBeRightAfter(C) \ {B}
                CanBeRightBefore(B) ← CanBeRightBefore(B) \ {C}
                if C∉MustBeBefore(B) then
                        add C«B
        if Valid(B)=1 then
            for each C∈MustBeAfter(B) s.t. Valid(C)≠0 do
                CanBeRightAfter(A) ← CanBeRightAfter(A) \ {C}
                CanBeRightBefore(C) ← CanBeRightBefore(C) \ {A}
                if C∉MustBeAfter(A) then
                        add A«C
    exit
```

Again, it is possible to show that if the precedence graph G is transitively closed (in the sense specified by Definition 1) and arc A « B is added to G then rule /2/ updates the precedence graph G to be transitively closed again. Note also, that propagation rules /1/ and /2/ achieve global consistency concerning the precedence constraint. This is a direct consequence of keeping a transitive closure of the precedence graph.

Proposition 3: If the precedence graph G is transitively closed (in the sense specified by Definition 1) and arc A « B is added to G then rule /2/ updates the precedence graph G to be transitively closed again.

Proof: Assume that arc A « B is added into G at a moment when arc B « C is already present in G. Moreover assume that Valid(A)≠0, Valid(B)=1, and Valid(C)≠0. We want to show that A « C is in G after rule /2/ is fired by the addition of A « B. The presence of arc B « C implies that C∈MustBeAfter(B) (and by symmetry also

B∈ MustBeBefore(C)). Now there are two possibilities. Either C∉ MustBeAfter(A) in which case rule /2/ adds the arc A « C into G, or C∈ MustBeAfter(A) (and by symmetry also A∈ MustBeBefore(C)) which means that arc A « C was already present in G when arc A « B was added.

The case when arc A « B is added into G at a moment when arc C « A is already present in G and Valid(C)≠0, Valid(A)=1, Valid(B)≠0 holds can be handled similarly. Thus when an arc is added into G, all paths of length two with a valid midpoint which include this new arc are either already spanned by a transitive arc, or the transitive arc is added by rule /2/. In the latter case this may invoke adding more and more arcs. However, this process is obviously finite (cannot cycle) as an arc is added into G only if it is not present in G, and no arc is ever removed from G. More on the time complexity of arc additions follows in Proposition 4.

Therefore, it is easy to see, that when the process of recursive arc additions terminates, the graph G is transitively closed. Indeed, for every path of length two in G with a valid midpoint one of the arcs on the path is added later than the other, and we have already seen that at a moment of such an addition the transitive arc is either already in G or is added by rule /2/ in the next step.   □

Proposition 4: The worst-case time complexity of the propagation rule /2/ (adding a new arc) including all recursive calls to rules /1/ and /2/ is $O(n^3)$, where n is the number of activities.

Proof: If arc A«B is added and B must also be before A then one of the activities A or B may become immediately invalid which takes time $O(n)$ (see Proof of Proposition 2). If both A and B are undecided then the rule prunes sets CanBeAfter(B) and CanBeBefore(A) and exits without further propagation. If A is valid and B is undecided (or vice versa) then all predecessors of A are connected to B. There are at most $O(n)$ such predecessors and the new arcs are added by recursive invocation of rule /2/. The recursion stops at this level because every predecessor X of a valid predecessor C of A is also a predecessor of A (due to the transitive closure) and hence the arc X«B has already been enqueued for propagation when addition of A«B was processed. Moreover, any duplicate copy of the same arc in the queue will be processed in time $O(1)$ (see the first line of rule /2/). The "worst" situation happens when both A and B are valid. Then all predecessors of A are recursively connected to all successors of B. There are at most $O(n^2)$ such connections and processing each connection takes time $O(n)$ – see the for loops in rule /2/, so the worst-case time complexity is $O(n^3)$.   □

Proposition 5: The rules /1/ and /2/ ensure that if B « A or there is a valid activity C between A and B (that is, A « C and C « B) then A∉ CanBeRightBefore(B) and B∉ CanBeRightAfter(A).

Proof: We will prove the proposition for the set CanBeRightBefore only, the set CanBeRightAfter is maintained symmetrically. At the beginning, the set CanBeRightBefore(B) contains all activities which is all right, because all activities are undecided. If A is deleted from CanBeBefore(B) (due to adding B « A), A is also deleted from CanBeRightBefore(B) in both rules /1/ and /2/. If any C becomes valid, A∈ MustBeBefore(C), and B∈ MustBeAfter(C) then A is deleted from CanBeRightBefore(B) in rule /1/. If a new arc A«C is added, C is valid, and

B∈ MustBeAfter(C) then A is deleted from CanBeRightBefore(B) in rule /2/. Similarly, if a new arc C«B is added, C is valid, and A∈ MustBeBefore(C) then A is deleted from CanBeRightBefore(B) in rule /2/.   ☐

4.3 A Propagation Rule for Direct Precedences

So far we more or less ignored the restrictions imposed by the state transition diagram. The reason is that these restrictions can be easily encoded by removing explicitly direct precedence relations from the double precedence graph. In particular, if transition from A to B is forbidden by the state transition diagram then arc A «$_d$ B is removed from the double precedence graph. In a totally ordered set of activities it implies that there must be some valid activity C between A and B or B must be after A. Actually a stronger requirement can be imposed: if A is before B (and A cannot be directly before B) then there must be some valid activity directly before B that is also after A and some valid activity directly after A that is before B. This observation can be transformed into the following implications:

$$CanBeRightAfter(A) \cap CanBeBefore(B) = \varnothing \implies B « A$$
$$CanBeAfter(A) \cap CanBeRightBefore(B) = \varnothing \implies B « A.$$

The above reasoning can be used to deduce a new precedence constraint B « A and, vice versa, if A « B then we can actively look for activities between A and B, especially, if there is only one candidate for such activity. This reasoning is realised using two propagation rules. First, the direct precedence is removed using rule /3/ and rule /4/ is activated. Rule /4/ is then called whenever there are some changes related to activities A or B. This rule tries to deduce that B must be before A or if A « B then the rule looks for some activity C between A and B.

```
A«dB is deleted  →                                           /3/
   CanBeRightAfter(A) ← CanBeRightAfter(A) \ {B}
   CanBeRightBefore(B) ← CanBeRightBefore(B) \ {A}
   activate rule /4/ for A and B
   exit

CanBeRightAfter(A) or CanBeAfter(A) or CanBeBefore(A) or
CanBeRightBefore(B) or CanBeBefore(B) or CanBeAfter(B) is changed, or
Valid(A) or Valid(B) is instantiated  →                      /4/
  if Valid(A)=0 or Valid(B)=0 or A∈MustBeAfter(B) then exit
  if CanBeRightAfter(A)∩CanBeBefore(B)=∅
      or CanBeAfter(A)∩CanBeRightBefore(B)=∅ then
        add B«A
        exit
  if A∈MustBeBefore(B) & Valid(A)=1 & Valid(B)=1 then
      if {C}=CanBeRightAfter(A)∩CanBeBefore(B) or
        {C}= CanBeAfter(A)∩CanBeRightBefore(B) then
            // C is the only possible direct successor of A or
            // C is the only possible direct predecessor of B
            add A«C
            add C«B
            Valid(C) = 1
            exit
```

If there are no explicit direct precedence relations like those imposed by the state transition diagram, then we already proved that propagation rules /1/-/2/ achieve global consistency. Unfortunately, global consistency cannot be achieved for rules /3/-/4/, that is, for explicitly removed direct precedence relations. Nevertheless, we can show that the constraint realised by rules /3/-/4/ is complete.

Proposition 6: If all activities are either valid or invalid and the set of valid activities is totally ordered then this order satisfies the restrictions imposed by the state transition diagram.

Proof: Assume for contradiction that there are valid activities A and B such that A is directly before B in the sequence but the state transition diagram forbids A to be directly before B. In such a case, rule /3/ has been called so B∉CanBeRightAfter(A) and rule /4/ is active. There is no invalid activity in CanBeRightAfter(A) due to rule /1/. For every valid activity C, either C « A or B « C and hence C∉CanBeRightAfter(A) due to rules /1/ or /2/. Recall that rule /4/ is called every time the set CanBeRightAfter(A) is changed. We just showed that CanBeRightAfter(A)=∅ and therefore also CanBeRightAfter(A)∩CanBeBefore(B)=∅. Therefore the second condition in rule /4/ is true and hence B « A is deduced which leads to failure. The rule /4/ cannot exit using the first condition because A and B are valid and A « B. The rule also cannot exit using the third condition because then there is a valid activity C such that A « C and C « B which is in contradiction with the order if activities. In any case, rule /4/ deduces failure so A cannot be right before B in any solution.   □

Proposition 7: The worst-case time complexity of the propagation rule /4/ including all recursive calls to rules /1/ and /2/ is $O(n^3)$, where n is a number of activities.

Proof: The time complexity of propagation rule /4/ alone is $O(n)$ because the intersection operations may require this time. The rule can add at most four arcs and it can make two activities valid. According to Proposition 2, making activity valid requires time $O(n^2)$. According to Proposition 4 adding an arc (including all recursive calls) requires time $O(n^3)$. Hence the total worst-case time complexity is $O(n^3)$.   □

5   Some Extensions

5.1   Sequence-Dependent Setup Times

The motivation for introducing direct precedence relations into precedence graphs was modeling sequence dependent setup times. Setup time is a time that must be inserted between two consecutive activities to setup the machine. If this setup time depends on both activities then it is called a *sequence-dependent setup time*.

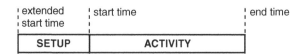

Fig. 3. Extending activity duration by setup time

Typically, setup time is assumed to be an empty gap between two consecutive activities. Our idea is to include the setup time in the duration of the second activity. Basically, it means that duration of each activity will consists of its real duration plus a setup time (Figure 3). Clearly because of time windows we still need to keep the original start time of the activity but we add the extended start time to model the start time including the setup. The extended start time can then participate in non-overlapping constraints for disjunctive resources such as edge-finding without modifying these constraints (the constraints only need support for variable duration).

The difference between the start time and the extended start time equals exactly to the setup time for a given activity (for the first activity, we can use a startup time). To find out the setup time we can use information about direct predecessors of the activity. Because we are working in the context of constraint satisfaction, we can use a binary constraint between the variable describing a direct predecessor (CanBeRightBefore) and the variable describing the setup time. The relation behind this constraint is extensionally defined – it is a list of setup times for the activity where index of each element corresponds to identification of the possible predecessor. Note finally, that these ideas can be included in filtering of times windows; we just need to assume variable duration of activities.

It may seem that a more natural model to describe setups is using setup activities inserted between any pair of activities in the state transition diagram. Thought this is theoretically possible, this approach has significant practical drawbacks. The main problem is decoupling the setup activity from the main activity. We need to describe the feature that if a "regular" activity is valid then exactly one of its predecessor setup activities must be valid. Hence we need additional logical constraints between the validity variables. Moreover, if the activities participate in other constraints, like edge-finding, this decoupling will decrease filtering power of these constraints (the constraints are not aware that the setup activity "belongs" to a regular activity). Finally, using setup activities will significantly increase the overall number of activities in the system because such a setup activity must be introduced for any pair of regular activities that may go directly one after another.

5.2 A Note on Open Graphs

The double precedence graphs studied in previous sections assume that the number of activities or at least its upper estimate is known. We use optional activities to deactivate activities that will not be part of the solution. This technique is appropriate in scheduling applications where most activities are known and optional activities are used to model alternatives to be decided during scheduling. However, in planning this technique is less convenient because the number of activities is unknown in general. It is still possible to use optional activities but in this case, the total number of activities will be probably too large which will decrease overall efficiency.

Our constraint model can be used directly to include new activities that will appear during problem solving. Recall, that we model the double precedence graph using difference sets, in particular the set CanBeBefore(A)\CanBeAfer(A) describes the activities that must be before A. We assumed that these sets are subsets of $\{1,\ldots,n\}$, where n is the number of activities. To model problems where the number of activities is unknown in advance, we can use an infinite set $\{1,\ldots,sup\}$, where sup is a computer representation of "plus infinity". The activities, that are already known, are represented using the variable Valid and sets CanBeBefore and CanBeAfter. The other void activities are represented just by their indices in these sets. Hence, these activities behave like optional undecided activities with no precedence relations to activities already in the graph. Therefore, there is no propagation related to these activities so sets representing these activities are not changing and hence it is not necessary to keep them in memory (only indices of invalid activities may be deleted from these sets, but it does not play any role). As soon as a new activity is included in the precedence graph then an index is assigned to the activity and its set representation is created. At this time all invalid activities should be removed from the sets of the new activity. We only need to keep the number of activities already included in the precedence graph to know which index can be used. Note finally, that we can still use optional activities to model alternatives to be decided later.

In addition to adding activities from outside, it is possible to use the double precedence graph to deduce that a new activity must be added to the graph. Notice that in case of open graphs, rule /4/ does actually no pruning because there can always be some void activities between A and B. Nevertheless, if A « B, A cannot be directly before B and no existing activity is between A and B then we can deduce that a new activity C must be added together with the precedence relations A « C and C « B. So we can take one of the void activities and make it real as described above. This technique might be useful especially to resolve flaws in plan-space planning.

6   Experimental Results

We implemented the model of the precedence graph with optional activities in SICStus Prolog 3.12.3 using the standard interface for the definition of global constraints. There are no independent benchmarks that include direct precedences so we present some preliminary experimental results comparing our approach with the constraint model from [5] using min-cutset problems. The min-cutset problem consists of precedence relations only and the task is to find the largest set of vertices such that the sub-graph induced by these vertices does not contain any cycle (or symmetrically to find the smallest set of vertices such that all cycles are broken if these vertices are removed from the graph). This problem is known to be NP-hard [7]. The original model uses a validity variable V and variable P indicating the absolute position of each activity in a sequence (its domain is $\{1,...,n\}$, where n is the number of activities). The precedence relation A « B is represented as a constraint $V_A * V_B * P_A < P_B$, which guarantees that there is no cycle between valid activities. Hence, to solve the problem, it is enough to decide the value of validity variables. In our model, we also need to decide only the value of validity variables; acyclic graph is guaranteed by keeping transitive closure. In both cases, the validity variables are instantiated by a standard backtracking algorithm with constraint propagation (branch-and-bound is used to maximize the number of valid activities).

We use the data set from [12] to compare both models. All the problems in the data set consist of 50 activities but the number of precedence constraints varies. Table 1 shows the maximal number of valid activities obtained for particular benchmarks. All the solutions obtained by our approach (Precedence) are optimal while the original

CLP model found and proved optimal solutions within the time limit of 50 minutes only for problems with 100 and 150 precedence constraints (the best solutions found within the time limit are presented for the remaining problems). The experiments run under Windows XP on 1.1 GHz Pentium-M processor with 1280 MB RAM.

Table 1. Min-cutset problems

Benchmark	P50-100	P50-150	P50-200	P50-250	P50-300	P50-500	P50-600	P50-700	P50-800	P50-900
Original	47	41	35	31	28	21	17	16	16	14
Precedence	47	41	37	33	31	22	19	17	16	14

Figure 4 shows the comparison of runtimes and the number of backtracks for both approaches. Our approach requires more than an order of magnitude less backtracks and less runtime to find the optimal solution. Note finally, that concerning the runtime we cannot compete with the GRASP heuristic proposed in [12], but this was not our original ambition as we tackle different problems. Moreover, opposite to the GRASP approach our technique is complete and, indeed, for some problems we have found better solutions than reported in [12].

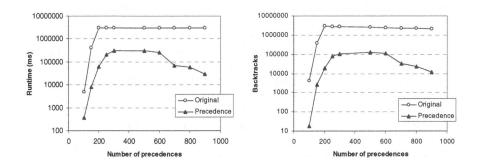

Fig. 4. Computation results on min-cutset problems (logarithmic scale)

7 Conclusion

We introduced a new constraint model describing precedence graphs with optional activities and direct precedence relations. For this model we proposed propagation rules that keep a transitive closure of the graph and remove inconsistencies caused by forbidden direct precedence relations. If explicit direct precedences are not present then the proposed rules achieve global consistency. We also experimentally showed that this model of the precedence graph is more efficient than a straightforward implementation of precedence relations. If explicit direct precedences, for example modelling state transition diagram, are present then the proposed rules realise a complete constraint model though the domain filtering is not complete. Rather than proposing a monolithic algorithm, we focused on incremental propagation of changes and on implementation-friendly architecture that is easy to translate into propagation rules usable in existing constraint solvers. Moreover this approach is extendable to problems where the number of activities in unknown in advance. Because the proposed technology is designed for resources with more complex behaviour, we believe that it might be appropriate for manufacturing scheduling with complex resources or for space applications such as scheduling earth observations.

Acknowledgments. The research is supported by the Czech Science Foundation under the contract no. 201/07/0205. We thank the anonymous reviewers for useful comments.

References

1. Baptiste, P., Le Pape, C., Nuijten, W.: Constraint-Based Scheduling: Applying Constraint Programming to Scheduling Problems. Kluwer Academic Publishers, Dordrecht (2001)
2. Barták, R.: Incremental Propagation of Time Windows on Disjunctive Resources. In: Proceedings of the Nineteenth International Florida Artificial Intelligence Research Society Conference (FLAIRS 2006), pp. 25–30. AAAI Press, Stanford (2006)
3. Beck, J.C., Fox, M.S.: Scheduling Alternative Activities. In Proceedings of AAAI-99. AAAI Press, USA (1999) 680–687
4. Cesta, A., Stella, C.: A Time and Resource Problem for Planning Architectures. In: Steel, S. (ed.) ECP 1997. LNCS, vol. 1348, pp. 117–129. Springer, Heidelberg (1997)
5. Fages, F.: CLP versus LS on log-based reconciliation problems for nomadic applications. In: Proceedings of ERCIM/CompulogNet Workshop on Constraints, Praha (2001)
6. Focacci, F., Laborie, P., Nuijten, W.: Solving Scheduling Problems with Setup Times and Alternative Resources. In: Proceedings of AIPS 2000, AAAI Press, Stanford (2000)
7. Garey, M.R., Johnson, D.S.: Computers and Intractability: A Guide to the Theory of NP-Completeness. W. H. Freeman and Company, San Francisco (1979)
8. Laborie, P.: Algorithms for propagating resource constraints in AI planning and scheduling: Existing approaches and new results. Artificial Intelligence 143, 151–188 (2003)
9. Laborie, P.: Resource temporal networks: Definition and complexity. In: Proceedings of the 18th International Joint Conference on Artificial Intelligence. IJCAI 2003, pp. 948–953 (2003)
10. Moffitt, M.D., Peintner, B., Pollack, M.E.: Augmenting Disjunctive Temporal Problems with Finite-Domain Constraints. In: Proceedings of the 20th National Conference on Artificial Intelligence (AAAI-2005), pp. 1187–1192. AAAI Press, Stanford (2005)
11. Stergiou, K., Koubarakis, M.: Backtracking algorithms for disjunctions of temporal constraints. In: Proceedings of the 15th National Conference on Artificial Intelligence (AAAI-98), pp. 248–253. AAAI Press, Stanford (1998)
12. Pardalos, P.M., Qian, T., Resende, M.G.: A greedy randomized adaptive search procedure for the feedback vertex set problem. Journal of Combinatorial Optimization 2, 399–412 (1999)

Reusing CSP Propagators for QCSPs

Marco Benedetti, Arnaud Lallouet, and Jérémie Vautard

University of Orléans
BP 6759 – F-45067 Orléans cedex 2
{Marco.Benedetti,Arnaud.Lallouet,Jeremie.Vautard}@univ-orleans.fr

Abstract. Quantified Constraint Satisfaction Problems are considerably more difficult to solve than classical CSP and the pruning obtained by local consistency is of crucial importance. In this paper, instead of designing specific consistency operators for constraints w.r.t each possible quantification pattern, we propose to build them by relying on classical existential propagators and a few analysis of some mathematical properties of the constraints. It allows to reuse a large set of constraints already carefully implemented in existing solvers. Moreover, multiple levels of consistency for quantified constraint can be defined by choosing which analysis to use. This can be used to control the complexity of the pruning effort. We also introduce QeCode, a full-featured publicly available quantified constraint solver, built on top of Gecode.

Keywords: QCSP, Quantified Languages.

1 Introduction

Quantified Constraint Satisfaction Problems (or QCSPs) are a generalization of classical Constraint Satisfaction Problems (CSPs) in which variables may be quantified universally and existentially [5]. This extension is promising in the sense that it allows to model problems that could not be modeled by CSPs like planning under uncertainty, games or model checking. But it also increases the complexity of solving problems, moving from NP-complete to PSPACE-complete.

In QCSPs, the classical notion of solution (an assignment of all variables that satisfies the constraints) is replaced by the more complex notion of *strategy* A strategy is a function which expresses the values taken by existentially quantified variables in function of the preceding universally quantified ones. For example, $\forall X \in \{1,2\} \exists Y \in \{1,2\} X = Y$ is true because we know that we can associate a value for Y to every value taken by X such that the constraint is true. A strategy is a complex object and its size is exponential in the number of variables.

Intuitively, we can liken a QCSP to a two-player game : one player (called the existential player) assigns the existentially quantified variables in order to satisfy all the constraints, while his opponent (called the universal player) assigns the universally quantified variables in order to violate at least one of the constraint. Each player assigns its variables in turns in the order they appear in the QCSP.

F. Azevedo et al. (Eds.): CSCLP 2006, LNAI 4651, pp. 63–77, 2007.
© Springer-Verlag Berlin Heidelberg 2007

constraint	condition	action
$\forall x \, \forall y \, \forall z \, (x + y = z)$	$x^- \neq x^+$ $\Longrightarrow$ false $y^- \neq y^+$ $\Longrightarrow$ false $z^- \neq z^+$ $\Longrightarrow$ false $x^- + y^- \neq z^-$ $\Longrightarrow$ false	
$\forall x \, \forall y \, \exists z \, (x + y = z)$	$z^- > x^- + y^-$ $\Longrightarrow$ false $z^+ < x^+ + y^+$ $\Longrightarrow$ false $\Longrightarrow I_z \subseteq [x^- + y^-, x^+ + y^+]$	
$\forall x \, \exists y \, \forall z \, (x + y = z)$	$z^- \neq z^+$ $\Longrightarrow$ false $z^- > x^- + y^-$ $\Longrightarrow$ false $z^+ < x^+ + y^+$ $\Longrightarrow$ false $\Longrightarrow I_y \subseteq [z^- - x^+, z^+ - x^-]$	
$\forall x \, \exists y \, \exists z \, (x + y = z)$	$x^- < z^- - y^+$ $\Longrightarrow$ false $x^+ > z^+ - y^-$ $\Longrightarrow$ false $\Longrightarrow I_y \subseteq [z^- - x^+, z^+ - x^-]$ $\Longrightarrow I_z \subseteq [x^- + y^-, x^+ + y^+]$	
$\exists x \, \forall y \, \forall z \, (x + y = z)$	$y^- \neq y^+$ $\Longrightarrow$ false $z^- \neq z^+$ $\Longrightarrow$ false $\Longrightarrow I_x \subseteq [z^- - y^+, z^+ - y^-]$	
$\exists x \, \forall y \, \exists z \, (x + y = z)$	$\Longrightarrow I_x \subseteq [z^- - y^-, z^+ - y^+]$ $\Longrightarrow I_z \subseteq [x^- + y^-, x^+ + y^+]$	
$\exists x \, \exists y \, \forall z \, (x + y = z)$	$z^- \neq z^+$ $\Longrightarrow$ false $\Longrightarrow I_x \subseteq [z^- - y^+, z^+ - y^-]$ $\Longrightarrow I_y \subseteq [z^- - x^+, z^+ - x^-]$	
$\exists x \, \exists y \, \exists z \, (x + y = z)$	$\Longrightarrow I_x \subseteq [z^- - y^+, z^+ - y^-]$ $\Longrightarrow I_y \subseteq [z^- - x^+, z^+ - x^-]$ $\Longrightarrow I_z \subseteq [x^- + y^-, x^+ + y^+]$	

Fig. 1. Propagation rules for the quantified version of the addition over intervals

The truth value of the problem corresponds then to the existance of a winning strategy for the existential player, i.e. the possibility for him to satisfy all the constraints, whatever its opponent does.

There is a growing interest in finding efficient techniques to solve such problems despite their complexity [5,2,6,4,7,8,3,10] but still the field is very young and for example no solver for quantified constraints is publicly available, and there is also no publicly available sets of benchmarks. This is to be put into perspective with the similar field of QBF where more than 15 solvers exist.

Quantified propagation will be at the heart of any efficient QCSP solver, if any, and the difficulty is that propagators are not the same for different patterns of quantifiers:

Example 1. Figure 1 reports quantified propagation rules for the ternary constraint $x + y = z$ evaluated over intervals. Rule 8 is the only thing a CSP reasoner might ever need to know (all the variables are existentially quantified). Conversely, a QCSP solver should also know as many of the other rules as possible to prevent the search procedure from carrying the burden of an almost-blind search. These rules are reported from [3], where they have been obtained through a case-by-case, manual analysis of the semantics of each operator against each quantification structure.

In this paper, we propose a contribution to the design of a solver for QCSPs. All existing approaches [5,6] have chosen to build the solver from scratch. This makes sense since the propagator for a constraint may be very different according to the quantification prefix associated to the constraint. But this solution is also very demanding in term of implementation effort and does not allow to reuse powerful constraint algorithms designed for the existential case. In what follows, we propose to study to what extent it is possible to reuse existing work for classical constraints in a quantified setting.

First, due to the properties of quantified propagation, the existential propagator is still correct. The only modification needed is to handle the case of reduction of the domain of an universal variable. Then we propose to add several analysis in order to prune more values: by using properties of the constraints like functionality, by using look-ahead or by extracting information from the dual. Theses analysis can be combined and for example, it is possible that a constraint hold no specific property but has a negation which is functional. Then a combination of dual and functional analysis may reveal interesting properties which yield to early discover an inconsistency or the pruning of some values.

2   QCSPs

In this section, we first recall the main definitions concerning QCSPs and set the notations. The definitions for QCSP we use are taken from [4], especially the elegant notion of *outcome* which allows to give a simple definition of quantified arc-consistency.

For a set D, we denote by $\mathcal{P}(D)$ its powerset and by $|D|$ its cardinality. Let V be a set of variables and $D = (D_X)_{X \in V}$ be the family of their finite domains. For $W \subseteq V$, we denote by D^W the set of tuples on W, namely $\Pi_{X \in W} D_X$. The projection of $A \subseteq D$ on $W \subseteq V$ is denoted by $A|_W$. If $Y \notin W$, the extension of a tuple $t \in D^W$ to $t' \in D^{W \cup \{Y\}}$ with $t'|_W = t$ and $t'_Y = a$ is denoted by $t : [Y = a]$. A *constraint* c is a couple (W, T) where $W \subseteq V$ are the variables of c (denoted by $var(c)$) and $T \subseteq D^W$ is the set of solutions of c (denoted by $sol(c)$).

Definition 2 (QCSP). *A* Quantified Constraint Satisfaction Problem *(or QCSP) is a 5-tuple* $\mathcal{Q} = (V, D, q, C)$ *where* $V = \{X_1, \ldots, X_n\}$ *is a linearly ordered finite set of variables,* $D = (D_X)_{X \in V}$ *is the family of their finite domains,* $q : V \to \{\forall, \exists\}$ *is a function which associates to each variable a quantifier and* C *is a set of constraints.*

For a QCSP $\mathcal{Q} = (V, D, q, C)$, we distinguish between its *quantification structure* (V, q) and its *CSP part* (V, D, C). In particular, we denote by QC a QCSP composed of a quantification structure Q and a CSP C on the same set of variables. A QCSP $\mathcal{Q} = (V, D, <, q, C)$ represents the closed formula $q(X_1) \in D_{X_1}, \ldots, q(X_n) \in D_{X_n} C$ where variables are numbered according to $<$. whose truth value can be given by recursive evaluation against each quantifier in turn.

The presence of a quantification structure changes the notion of solution with respect to classical CSPs. Indeed, a solution of a QCSP is no more described by the assignment of a value to the different variables but by the more general notion of *strategy* in which every existentially quantified variable is given a value in function of the preceding universally quantified ones (called Skolem functions). For a quantification structure $Q = (V, <, q)$, we denote respectively by $V[Q, \forall]$ and $V[Q, \exists]$ the set of universally and existentially quantified variables of Q. For $X \in V$, $V[Q, < X]$ denotes the set of variables before X in the order $<$. Combinations are allowed and $V[Q, \forall, < X]$ denotes the set of universal variables preceding X in Q. Formally, a *strategy* s for a QCSP QC is a family of functions $(s_X)_{X \in V[Q, \exists]}$ such that $s_X : D^{V[Q, \forall, < X]} \to D_X$.

Let s be a strategy. A *scenario* is the V-tuple formed by one branch of s, i.e. the assignment of each $\forall$-variable with one of its domain value and the corresponding evaluation of the Skolem functions for each $\exists$-variable. The set of scenarios of a strategy s is denoted $sce(s)$. A strategy is a *winning strategy* if all of its scenarios satisfy all the constraints of C. The set of all winning strategies for a QCSP QC is denoted by $WIN(QC)$. A strategy can be described as a tree enumerating all possibilities for universally quantified variables and giving the value of the existentially quantified ones for every possibility.

QCSPs and CSPs are close enough to allow to reuse most of the technology, including local consistency. A value for an $\exists$-variable does not have to be tested if it does not belong to any winning strategy. In a similar way, if a value for an $\forall$-variable does not belong to any winning strategy, then the problem has no solution. These properties can be tested in some cases by local consistency, although it provides only sufficient conditions for them. In order to define the counterpart of arc-consistency in the quantified case, it is useful to define the set $out(QC)$ of *outcomes* of a QCSP as the union of all scenarios of all winning strategies. A value is globally inconsistent if it does not appear in any outcome. But finding this requires to solve the whole QCSP, making this information pointless. Let us call *quantified constraint* a QCSP formed only with one single constraint c. If a value is *locally inconsistent*, that is it does not appear in any outcome of Qc, then it is also globally inconsistent. Thus it is possible to devise propagator for quantified constraints which enjoy many of the good properties of the classical ones [3].

A *search state* is a set of yet possible values for each variable: for $W \subseteq V$, it is a family $s = (s_X)_{X \in W}$ such that $\forall X \in W, s_X \subseteq D_X$. The corresponding *search space* is $S_W = \Pi_{X \in W} \mathcal{P}(D_X)$. A consistency can be modeled as the greatest fixpoint of a set of so-called *propagators* and is computed by a chaotic iteration [1]. For a constraint $c = (W, T)$, a *propagator* is an operator f on S_W[1] having the following properties:

- *monotonicity*: $\forall s, s' \in S_W, s \subseteq s' \Rightarrow f(s) \subseteq f(s')$.
- *contractance*: $\forall s \in S_W, f(s) \subseteq s$.
- *correctness*: $\forall s \in S_W, \Pi s \cap sol(c) \subseteq \Pi f(s) \cap sol(c)$.
- *singleton completeness*: for all $s \in S_W$ such that $|\Pi s| = 1$,
$$\Pi f(s) \cap sol(c) \subseteq \Pi s \cap sol(c).$$

---

[1] When iterating operators for constraints on different sets of variables, a classical cylindrification on V is applied.

The purpose of a propagator is to remove any value which does not occur in any solution. For $c = (W, T)$, the well-known arc-consistency operator ac_c is defined by: $\forall s \in S_W, ac_c(s) = s'$ with $\forall X \in W, s'_X = (\Pi s \cap T)|_X$. *Quantified arc-consistency* ([5] and called strong quantified GAC in [8]) can be defined as a propagator: $\forall s \in S_W, qac_c(s) = s'$ with $\forall X \in W, s'_X = (\Pi s \cap out(Qc))|_X$. Obviously, a way to perform such a pruning is to solve the constraint as a little QCSP by enumerating all possible values in the domain of the variables. We propose in what follows to study what can be done with less computational efforts. In this case, the pruning may of course be weaker.

The way universally quantified variables are handled differs from existential ones: whenever its domain is reduced, the whole problem has no solution and inconsistency can be immediately derived.

3 Analysis

Quantified arc-consistency is powerful but costly. Indeed, it is not always possible to enforce it without actually trying all possible values for the variables: the cost of one exploration for a n-ary constraint with a domain size of d is at most d^n [2]. This corresponds to the development of the search tree on the variables of this constraint. Thus computing quantified arc-consistency for a global constraint may be counter-productive as it would require to develop a large enumeration tree that would be redundant with respect to the main search tree. Hence, the control of the level of consistency to be applied is even more important than in classical CSPs.

In this section, we describe four kinds of analysis that can be performed in order to prune domain values of quantified variables. Some of these techniques amount to perform quantified arc consistency on a relaxed version of the quantified constraint as shown by the following remark:

Remark 3. Let Qc be a quantified constraint and $A \subseteq D^V$ be such that $out(Qc) \subseteq A$. Then the operator f defined by $f_X(u) = (u \cap A)|_X$ is correct.

Such a set A can be obtained as $out(Q'c')$ for a specific relaxation $Q'c'$ of Qc.

3.1 Existential Analysis

Existential analysis consists in applying the classical arc-consistency operator (quantifiers are simply ignored). The only difference resides in the treatment of universally quantified variables. Indeed, whenever the domain of an universal variable is reduced, the problem is insatisfiable. Hence existential variables are reduced and universal ones are used to early detect inconsistency.

In order to prove correctness of the relaxation, we introduce the following order between two quantification structures $Q = (V, <, q)$ and $Q' = (V, <, q')$ having the same set of variables and order: $Q \subseteq Q'$ iff $\forall X \in V, q(X) = \exists \rightarrow q'(X) = \exists$. It means that some universal quantifiers may be replaced by existential ones.

Theorem 4. *If $Q \subseteq Q'$, then $out(Qc) \subseteq out(Q'c)$.*

Proof. Let Q and Q' be two quantification structures such that $Q \subseteq Q'$. The proof is by induction on the number of different quantifiers beetween Q and Q'. If $Q = Q'$, we immediately have $out(Qc) = out(Q'c)$. Suppose now that Q and Q' differ only by one quantifier, i.e. there exists a unique variable Y such that Y is universally quantified in Q and existentially quantified in Q'. If Qc is false, we immediately have $out(Qc) = \emptyset \subseteq out(Q'c)$. If Qc is true, then we have to prove that every outcome of Qc is an outcome of $Q'c$, i.e. that it is a scenario of a winning strategy s' of Q'.

Let o be an outome of Qc. It belongs to a winning strategy s of Qc. Let a be the value given to Y by o. We have $s = (f_X)_{X \in V[Q,\exists]}$ with f_X being the Skolem function taking in argument the universal variables preceding X. We build a strategy s' of $Q'c$, consisting of the following Skolem functions $(f'_X)_{X \in V[Q',\exists]}$ such that:

- $\forall X \in V[Q',\exists,<Y], f'_X = f_X$
- $f'_Y : D^{V[Q',\forall,<Y]} \to D_Y$ is the constant function a
- otherwise $f'_X : D^{V[Q',\forall,<X]} \to D_X$ maps the tuple t onto the value $f_X(t : [Y = a])$.

Clearly, $o \in sce(s')$ since it gives the same values to every variable. The property is also true for every scenario of s, (it is possible to build a strategy s' such that $o \in sce(s')$). For any scenario p' of s', the value assigned to Y is a and by construction, p' is also a scenario of s. Since s is a winning strategy, s' is also a winning strategy and we have $o \in out(Q'c)$. By induction on the number of different quantifier between Q and Q', we can deduce this result for every Q and Q' such that $Q \subseteq Q'$. $\square$

We define the following operator which reduces a quantified constraint according to its existential relaxation:

Definition 5. *Let Qc be a quantified constraint and $Q' = \max\{G \mid Q \subseteq G\}$. We call* existential consistency *the operator $eac(Qc) = qac(Q'c)$.*

Note that existential concistency is strictly more powerful than usual arc-consistency of CSPs, as it returns $\emptyset$ if the domain of a universal variable have been touched.

Any value suppressed by this propagator would also be suppressed by quantified arc-consistency:

Corollary 6. *The propagator $eac(Qc)$ is correct.*

Proof. It follows from Remark 3 and Theorem 4. $\square$

It was known in the literature [3] that quantified consistency is stronger than existential consistency, in the sense that couples variable/value that appear in no solution strategy of a quantified problem (and could be safely removed) can be consistent with the same problem when all the variables are considered as existentially quantified (hence they are not removed). We provide here a proof of this (much expected) result, that was apparently missing in the original papers.

Example 7. For example, consider the quantified constraint $\exists X \in \{1,2\}\ \forall Y \in \{1,2\}\ \exists Z \in \{1,2,3,4,5\}\ X + Y = Z$. The existential propagator reduces Z's domain to $\{2,3,4\}$. Also, if the quantified constraint was as follows: $\exists X \in \{1,2\}\ \forall Y \in \{1,2\}\ \exists Z \in \{1,2\}\ X + Y = Z$, the existential propagator would have removed 2 in the domain of Y, thus detecting inconsistency.

Though incomplete, existential consistency is a correct relaxation of quantified consistency, so that its application is always safe. But it is incomplete with respect to quantified arc-consistency, as shown by the following example:

Example 8. Consider the quantified constraint $\exists X \in \{1,2\}\ \forall Y \in \{1,2\}\ \exists Z \in \{2,3\}\ X + Y = Z$. The existential propagator does not reduce X's domain. Also, if the quantified constraint was as follows: $\exists X \in \{1,2\}\ \forall Y \in \{1,2\}\ \forall Z \in \{2,3,4\}\ X + Y = Z$, the existential propagator would not have detected inconsistency.

However, existential analysis is surprisingly sufficient for binary constraints. Binary constraints may have four different quantification structures: $\forall\forall$, $\forall\exists$, $\exists\forall$ and $\exists\exists$. Existential analysis performed on the last case is classical arc-consistency. As mentioned in [7], $\forall\forall$ and $\exists\forall$ can be removed statically. The first case is only true for the true constraint. In the second case, every values of the existential variable that is not supported by all values of the universal one can be removed and at the end, the constraint is entailed and thus can be removed from the problem. For the remaining case, we can show the following result:

Proposition 9. *Let* $\forall X\ \exists Y\ c(X,Y) = Qc$ *be a quantified constraint. Then* $qac(Qc) = eac(Qc)$.

Proof. We only need to prove that $eac(Qc) \subseteq qac(Qc)$. The other direction is obtained by Corollary 6. We simply remark that a value in the domain of Y supressed by qac is also supressed by eac and an inconsistency caused by a reduction of X's domain is also detected by eac.   □

This result is interesting because it shows that quantified arc-consistency is not costly for binary constraints. Moreover, this property can be generalized to any n-ary quantified constraint in the following way. Let us consider the following order on quantification structures $Q = (V, <, q)$ and $Q' = (V, <', q)$ having the same set of variables and the same quantification function: we have $Q \sqsubseteq Q'$ iff $\forall X, Y \in V, (X < Y \wedge Y <' X) \rightarrow q(Y) = \forall$. In other words, Q and Q' differ in that some universally quantified variables have moved to the left. For example, $\forall X\ \exists Y\ \forall Z \sqsubseteq \forall X\ \forall Z\ \exists Y$.

Theorem 10. *If* $Q \sqsubseteq Q'$, *then* $out(Qc) \subseteq out(Q'c)$.

Proof. Let Q and Q' be two quantification structures such that $Q \sqsubseteq Q'$. The proof is by induction on a path from Q to Q' in which only one quantifier moves at each step. So, suppose that variable X_j moves before X_i, the order in Q is $X_1, \ldots, X_i, \ldots, X_j, \ldots, X_n$ and the order in Q' is $X_1, \ldots, X_{i-1}, X_j, X_i, \ldots, X_n$. By definition, we have $q(X_j) = q'(X_j) = \forall$.

Let $o \in out(Qc)$ be an outcome such that a is the value given to V_j by o. Then, it belongs to a winning strategy s of Qc. We build a winning strategy s' of $Q'c$ such that $o \in sce(s')$:

- $\forall X \in V[Q', \exists, < V_j], f'_X = f_X$
- otherwise, f'_X is defined by $f'_X(t) = f_X(t : [V_j = a])$.

Clearly, $o \in sce(s')$ since it gives the same values to every variable. The property is also true for every scenario of s, (it is possible to build a strategy s' such that $o \in sce(s')$). For any scenario p' of s', the value assigned to Y is a and by construction, p' is also a scenario of s. Since s is a winning strategy, s' is also a winning strategy and we have $o \in out(Q'c)$. By induction on a path of one-quantifier moves, we have $out(Qc) \subseteq out(Q'c)$. $\square$

This theorem can be used to provide a relaxation for quantified constraints. When only one universal quantifier is before existential ones, quantified arc-consistency can be computed by using classical arc-consistency:

Proposition 11. *Let Q be a quantification structure such that only the first variable is quantified universally, then $qac(Qc) = eac(Qc)$.*

Proof. let c be a n-ary constraint on variables $X_1, \ldots, X_n$ and Q the quantification structure making X_1 universal and all other variables existential. We only need to prove that $eac(Qc) \subseteq qac(Qc)$. Suppose that a value b has not been removed from the domain of an existential variable Y in eac and is removed in qac. Then there exists a tuple $t \in sol(c)$ such that $t|_Y = b$. With this tuple, we can build a strategy such that $f_Y(t|_X) = b$ (other cases being unchanged). It means that $t \in out(Qc)$, which contradicts the hypothesis of its removal by qac. If we suppose that a value a has not been removed from the (only universal) variable X_1 by eac, but has been removed by qac, then it means that there is both a tuple and no tuple supporting this value, hence the contradiction. Consequently, we have $eac(Qc) = qac(Qc)$. $\square$

3.2   Functional Domain Analysis

In this section we show how to infer that a quantified constraint is inconsistent by just reasoning on some matematical properties of the constraint itself, and on their relation to the size of the variable domains. Namely, we show how the interaction between the quantification structure of the problem and the functional dependencies that are possibly embedded in a constraint might lead to an inconsistency.

Definition 12 (Functional Constraint). *An n-ary constraint c is functional on its i-th argument iff every time that $c(a_1, \ldots, a_i, \ldots, a_n)$ and $c(b_1, \ldots, b_i, \ldots, b_n)$ with $a_i \neq b_i$ then necessarily $a_j \neq b_j$ for some $j \neq i$.*

The intuition behind this definition is that the relation defining the constraint is actually a function from the rest of the arguments onto the i-th one. If this function happens to be *injective*, we say the constraint is *injectively functional*. The property of being functional is a feature of the constraint itself, and is not dependant on the particular domain sets we consider (though a constraint that in general is not functional can behave as a functional one on specially shaped domains). So, each argument of each constraint needs to be labeled as functional or non-functional just once and for all, at the time the constraint is defined.

A constraint can be functional on one, some, or all of its arguments. For example, the ternary constraint $x + y = z$ is functional on every argument (but is not injective), $x \times y = z$ is only functional on z, $x \leq y$ is not functional, and $y = 2x$ is injectively functional on both arguments.

Given a functional constraint, we might be able to detect an "interference" (between the functional dependencies in the quantification structure and the functions in the constraint) that necessarily that generates an inconsistency when we take into account the size of the domains.

Example 13. Consider a fragment of a quantified problem like:

$$\forall y.\exists x.C(x, y)$$

where C is functional on y. This means that y is a function of x in the constraint, and that, given the prefix, x is a function of y in the whole QCSP: There is a reciprocal functional dependency between variables with different quantifiers, where the existential one is deeper. Each time we choose a value for y we need a value for x that makes the constraint true. But, once the value of x is found we can associate to it a *unique* value of y via C. Suppose we are in the lucky circumstance to always observe a consistent mutual association (if not, the constraint is violated). Still, we are forced to associate a different value of x to every value of y (we *cannot* reuse the same twice because $C : x \to y$ is a function). If we run out of x values before the y values have been exhausted—i.e. if $|D(x)| < |D(y)|$—we have a sufficient (not necessary) condition of inconsistency.

This is a special case of a more general condition captured by the following result.

Theorem 14 (Functional Inconsistency). *A sufficient condition for a quantified constraint $Qc(X_1, \ldots, X_n)$ to be inconsistent is:*

1. *The constraint is functional on X_i*
2. *$q(X_i) = \forall$*
3. *$(\prod_{X_i \prec X_j, q(X_j)=\exists} |D(x_j)|) < |D(x_i)|$*

An interesting special case of the former theorem one encounters when the universal functional argument appears as an independent variable in no prefix function, i.e. when it is to the right of every existentially quantifiers involved in the constraint. In this case, the last condition becomes $|D(x_i)| > 1$: if the variables not-yet-assigned in the left-to-right order of the prefix when x_i is encountered are all universals, then the consistent value for x_i (if any) is functionally specified, hence it is unique. If the domain of x_i contains more than one element, the constraint can be made false.

Example 15. Let us consider how a functional analysis (jointly with an existential analysis) can simulate the first rule in Table 1. If $x^- \neq x^+$ then $|D(x)| > 1$. But, the constraint is functional in x, and no existential quantifier to the left of x exists (actually, no existential quantifier at all is mentioned). So, $|D(x)| > 1$ is sufficient to infer the inconsistency. A similar reasoning holds for the cases $y^- \neq y^+$ and $z^- \neq z^+$. If no inconsistency is detected after this functional analysis, each domain necessarily has cardinality equal to one. In this case, the $x^- + y^- \neq z^-$ check is equivalent to the existential analysis.

An injectively functional constraint is a functional constraint, hence subject to the sufficient conditions discussed above, but its additionally property can be exploited.

Theorem 16 (Injectively Functional Inconsistency). *A sufficient condition for a quantified constraint $Qc(X_1, \ldots, X_n)$ to be inconsistent is:*

1. *The constraint is injectively functional on X_i*
2. $(\prod_{X_i \prec X_j, q(X_j)=\forall} |D(x_j)|) > 1$

We notice that the inference power of functional domain analysis is incomparable with the rules given in Table 1. We have already encountered a case in which those rules are strictly more powerful than domain analysis (if the domain contains holes). However, it is easy to show examples where the opposite holds.

Domain functional analysis can be generalized to *sets* of functional constraints (that share some variable), yielding a strictly more powerful rule.

Example 17. Inconsistency *cannot* be inferred by functionally analyzing each single constraint, nor even by existential analysis, in

$$\exists z \in \{1, 2\} \; \forall y \in \{4, 16\} \; \exists x \in \{2, 4\}$$
$$x^2 = y \wedge 2z - x = 0$$

However, y is a function of x in the first constraint, and x is a function of z in the second. So, a functional domain analysis for the overall constraint $C' : z \to y$ proves the inconsistency, because $|D(y)| > 1$ and z preceeds y.

3.3   Look-Ahead Analysis

Look-ahead consists in applying an enumeration step to a variable of the constraint. If the enumerated variable is universal, the reduced domains are combined with intersection and if it is existential, with union. Look-ahead has to be combined with a specific consistency and we assume for now that it is the existential one. For $c = (W, T)$, we can then define the look-ahead of Qc on the existential variable $X \in W$ such that, on a given search state s, $l(X)(s) = s'$ with $s'_Y = \bigcup_{a \in D_X} eac_c(s[X \leftarrow a])|_Y$. Similarly, the look-ahead of Qc on the universal variable $X \in W$ is $l(X)(s) = s'$ with $s'_Y = \bigcap_{a \in D_X} eac_c(s[X \leftarrow a])|_Y$. Quantified arc-consistency can be computed by a look-ahead on every variable in the quantification order.

Look-ahead can be particularly efficient for convex constraints:

Definition 18 (Convex Constraint). *An n-ary constraint c over ordered sets is* convex *on i-th argument iff every time that $c(a_1, \ldots, a_{i-1}, a_i, a_{i+1}, \ldots, a_n)$ and $c(a_1, \ldots, a_{i-1}, b_i, a_{i+1}, \ldots, a_n)$ with $a_i < b_i$ then necessarily $c(a_1, \ldots, a_{i-1}, x, a_{i+1}, \ldots, a_n)$ holds for every $a_i < x < b_i$. The constraint is simply* convex *if it is convex on every argument.*

The class of convex constraints contains interesting elements, such as the widely used "linear constraints" that express some disequality relation between linear expressions, e.g.

$$x + 2y \leq 3z$$

For convex constraints we can prove a simple but useful property. We call *convex look-ahead* a consistency check in which only the bounds of a variables are tested.

Theorem 19 (Convex Lookahead). *Given a convex constraint $\mathcal{Q}_1 x_1 \ldots \mathcal{Q}_n x_n$. $C(x_1, \ldots, x_n)$, the n-step look-ahead procedure detects an inconsistency iff the n-step convex look-ahead does.*

Example 20. A 1-step convex lookahead is enough to simulate the $\forall x \forall y \exists z$ rule in Table 1. We already know that the contraction of the domain of z can be realized by existential analysis, while the same kind of analysis might fail to capture the three inconsistency conditions. For example, under $z^- > x^- + y^-$ the value x^- is removed by existential analysis (hence the inconsistency is detected) only if it is also $z^- < x^- + y^+$. But if we consider the situation

$$\forall x \in \{0, 1\} \forall y \in \{1, 2, 3\} \exists z \in \{2, 3\}.x + y = z$$

the existential analysis says nothing. A functional analysis may provide enough inference power to infer inconsistence, like in this case (the constraint is functional on y, and y dominates the existential variable z in the prefix, and $|D(z)| < |D(y)|$). However, the lifted case

$$\forall x \in \{0, 1\} \forall y \in \{1, 2, 3\} \exists z \in \{2, 3, 4\}.x + y = z$$

is still FALSE but this time subject to no dimensional inference. A one-step lookahead on the x^- value solves the problem : at the first step, by taking value 0 for x, we obtain:

$$\forall y \in \{1, 2, 3\} \exists z \in \{2, 3, 4\}.y = z$$

which is existentially reduced to an inconsistency by eliminating 1 from the $D(y)$. More in general, the two 1-step sub-problems identified in the convex constraint analysis on x or y solve the problem in case the previous techniques do not apply.

3.4 Dual Analysis

By *dual analysis* we mean any form of inference that takes into account at the same time the *direct* QCSP problem, and its *dual* version. The dual version is obtained by inverting each quantifier in the prefix, and negating the set of constraints. The dual problem is true if and only if the direct one is false.

Dual reasoning in the framework of QCSP has been proposed for the first time in [2], where it was used to help the search procedure to be less redundant. Indeed, the dual problem models winning strategies for the universal opponent. When a value can be removed from an existential domain in the dual problem, the same value can be safely *ignored* in the domain of the corresponding universal variable of the direct problem, because it is guaranteed to make the (dual) problem false, hence it does not contradict any constraint in the (direct) problem. The overall effect is that of reducing the branching factor during universal branching, because a smaller set of cases is to be considered.

The way we use dual reasoning in building a QCSP solver is slightly different. Essentially, we exploit the analysis of the dual problem as a way to strengthen the inference power of a purely existential propagator.

Example 21. Let us consider the problem

$$\forall x \in \{1, 2\} \forall y \in \{1, 2\} \exists z \in \{1, 2\}.C(x, y, z)$$

and its dual problem

$$\exists x \in \{1, 2\} \exists y \in \{1, 2\} \forall z \in \{1, 2\}.\overline{C}(x, y, z)$$

where the constraints C and $\overline{C}$ are explicitly defined as

$C(x, y, z)$				$\overline{C}(x, y, z)$		
x	y	z		x	y	z
1	2	1		2	1	1
2	1	2		2	2	1
1	1	1		2	2	2
1	1	2				
1	2	2				

Even though the problem is false, the existential analysis on $C(x, y, z)$ is able to infer nothing (no variable domain can be reduced). However, if we perform an existential analysis on $\overline{C}(x, y, z)$ we are able to reduce the (existential) domain of x from $\{1, 2\}$ to $\{2\}$. This means that for $x = 1$ the dual problem is false, hence the direct problem is true. So, we can reduce the interesting cases for the domain of the universal variable x in the direct problem to $\{2\}$ only (in the other cases we know that the universal player cannot win). Given that only the tuple $\langle 2, 1, 2 \rangle$ remains under this restriction, we can apply existential reasoning again in the direct problem and reduce the domain of y from $\{1, 2\}$ to $\{2\}$, hence we derive an inconsistency.

As the example shows, dual analysis may be fruitfully employed during consistency check for a single constraint, not necessarily for the problem as a whole. Still, the most interesting case would be to reason on a completely mirrored problem. While, as discussed in [2], it is true that *"there is no theoretical obstacle in obtaining the negated version of the problem in a finite-domain setting"*, it has to be taken into account that the negation of a conjunction of constraints is a *disjunction of negated constrains*. This fact poses two challenges to any CSP-based solution willing to re-use existing solutions:

- Some of the negated constraints may have a quite natural, one-constraint representation in the system (e.g. we move from $x > y + z$ to $x \leq y + z$), others may not (e.g. the negation of a disjunction).
- Furthermore, any CSP system is built to reason on *conjunctions* of constraints. The fixpoint computation for propagators and the management of domain variables are seriously affected by the shift to the disjunctive form.

4 Experiments

An implementation of a QCSP solver following these principles is in progress on top of the solver Gecode [9]. It allows to reuse all existing constraints of Gecode, including global constraints, for which the existential propagator is provided. We are aware of no other QCSP solver taking as input a constraint language as expressive as the one we consider. So, we compare our implementation on a greatly restricted input language, for which benchmarks and solvers exist though. Namely, we refer to the random model generator described in [7], and compare with the state-of-the-art results reported therein. That model only generates QCSPs over *binary constraints* whose truth tables have been *explicitly* provided. Hence, the very core of our work is not exercised. At least, the comparison gives meaningful indication on the level of integration that has been achieved between our quantified propagation and search scheme and the robust and publicly available CSP solver GeCode.

Figure 2 present results for two settings of the random generator. All the data but the QeCode's ones are taken from [7]. In the left picture, the mean average

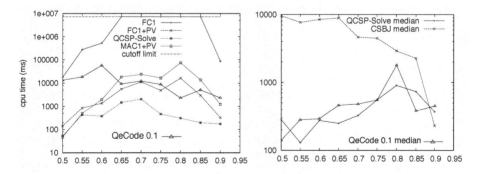

Fig. 2. Comparison over the random instances described in [7]. In the left picture the parameters are $n = 21$, $n_\forall = 7$, $d = 8$, $p = 0.2$, $q_{\forall\exists} = 1/2$. In the right picture it is the same but $n = 24$, $n_\forall = 8$, $d = 9$. The value of the $q_{\exists\exists}$ parameter is given on the x axis.

value of 100 runs for each value on the x axis is reported, and the comparison is against the state of the art QCSP-Solve solver[2] in various configurations (obtained by switching on/off advanced search/learning techniques). In the right picture, the median value of 100 runs is compared against the best QCSP-Solve configuration and a QBF-based approach called CSBJ.

The relatively good experimental behavior of QeCode is extremely promising for the following reasons. First, QCSP-Solve focuses on implementing several enhancements to the search procedure which are not key to this paper and are thus missing in QeCode 0.1 (namely: search-level look-ahead, pure-value detection, conflict-directed backjumping, solution-directed pruning, and symmetry detection). Such techniques will be added in future releases. Second, QCSP-Solve and CSBJ have been written from scratch with QCSP problems in mind, while the QeCode way of solving the above random models is really a *cooperation* between GeCode's machinery and our techniques. For example, the search is driven by QeCode, but the fixpoint computation of propagation is done by GeCode, and during such propagation QeCode intervenes with special propagators that simulate the effect of universal quantification. Last but absolutely not least, QCSP-Solve and CSBJ are tailored for explicitly defined binary constraints (and are only able to handle that ones) while QeCode can take as input any constraint GeCode can, and explicitly defined binary ones are just a very special case. In the current status, existential and functional analysis are provided, as well as a SAT-inspired implementation of heuristics. QeCode will be shortly available, and accessible from the authors web page.

5   Conclusion

In this paper, we propose a conservative scheme to reuse a classical CSP solver in the quantified case. It requires several analysis of the constraints: existential, functional, dual analysis and look-ahead. Even without look-ahead which could yield full quantified arc-consistency, these techniques are shown to be remarkably powerful on a first version of QeCode, a quantified solver built on top of GeCode.

References

1. Apt, K.R.: The essence of constraint propagation. Theoretical Computer Science 221(1-2), 179–210 (1999)
2. Bordeaux, L.: Résolution de problèmes combinatoire modélisés par des contraintes quantifiées, PhD thesis (2003)
3. Bordeaux, L.: Boolean and interval propagation for quantified constraints. In: Gent, I., Giunchiglia, E., Stergiou, K. (eds.) Workshop on Quantification in Constraint Programming, 2005, Barcelona, Spain (2005)
4. Bordeaux, L., Cadoli, M., Mancini, T.: CSP properties for quantified constraints: Definitions and complexity. In: Veloso, M.M., Kambhampati, S. (eds.) National Conference on Artificial Intelligence, pp. 360–365. AAAI Press, Stanford (2005)

---

[2] As already said, this solver only takes as input explicitly defined binary constraints.

5. Bordeaux, L., Monfroy, E.: Beyond NP: Arc-consistency for quantified constraints. In: Hentenryck, P.V. (ed.) CP 2002. LNCS, vol. 2470, pp. 371–386. Springer, Heidelberg (2002)
6. Gent, I.P., Nightingale, P., Rowley, A.: Encoding quantified CSPs as quantified boolean formulae. In: Mántaras, R.L.d., Saitta, L. (eds.) ECAI, pp. 176–180. IOS Press, Amsterdam (2004)
7. Gent, I.P., Nightingale, P., Stergiou, K.: QCSP-Solve: A solver for quantified constraint satisfaction problems. In: Kaelbling, L.P., Saffiotti, A. (eds.) IJCAI, pp. 138–143. Professional Book Center (2005)
8. Nightingale, P.: Consistency for quantified constraint satisfaction problems. In: van Beek [11], pp. 792–796
9. Schulte, C., Tack, G.: Views and iterators for generic constraint implementations. In: van Beek [11], pp. 817–821
10. Stergiou, K.: Repair-based methods for quantified csps. In: van Beek [11], pp. 652–666
11. van Beek, P. (ed.): CP 2005. LNCS, vol. 3709, pp. 1–5. Springer, Heidelberg (2005)

Bipolar Preference Problems:
Framework, Properties and Solving Techniques

Stefano Bistarelli[1,2], Maria Silvia Pini[3], Francesca Rossi[3], and K. Brent Venable[3]

[1] Dipartimento di Scienze, Università "G. d'Annunzio", Pescara, Italy
bista@sci.unich.it
[2] Istituto di Informatica e Telematica, CNR, Pisa, Italy
Stefano.Bistarelli@iit.cnr.it
[3] Dipartimento di Matematica Pura ed Applicata, Università di Padova, Italy
{mpini,frossi,kvenable}@math.unipd.it

Abstract. Real-life problems present several kinds of preferences. We focus on problems with both positive and negative preferences, that we call *bipolar preference problems*. Although seemingly specular notions, these two kinds of preferences should be dealt with differently to obtain the desired natural behaviour. We technically address this by generalizing the soft constraint formalism, which is able to model problems with one kind of preferences. We show that soft constraints model only negative preferences, and we define a new mathematical structure which allows to handle positive preferences as well. We also address the issue of the compensation between positive and negative preferences, studying the properties of this operation. Finally, we extend the notion of arc consistency to bipolar problems, and we show how branch and bound (with or without constraint propagation) can be easily adapted to solve such problems.

1 Introduction

Many real-life problems contain statements which can be expressed as preferences. Our long-term goal is to define a framework where many kinds of preferences can be naturally modelled and efficiently dealt with. In this paper, we focus on problems which present positive and negative preferences, that we call *bipolar preference problems*.

Positive and negative preferences can be thought as two symmetric concepts, and thus one can think that they can be dealt with via the same operators. However, this would not model what one usually expects in real scenarios. In fact, usually combination of positive preferences should produce a higher (positive) preference, while combination of negative preferences should give us a lower (negative) preference.

When dealing with both kinds of preferences, it is natural to express also indifference, which means that we express neither a positive nor a negative preference over an object. Then, a desired behaviour of indifference is that, when combined with any preference (either positive or negative), it should not influence the overall preference.

Finally, besides combining preferences of the same type, we also want to be able to combine positive with negative preferences. We strongly believe that the most natural and intuitive way to do so is to allow for compensation. Comparing positive against negative aspects and compensating them w.r.t. their strength is one of the core features

F. Azevedo et al. (Eds.): CSCLP 2006, LNAI 4651, pp. 78–92, 2007.

of decision-making processes, and it is, undoubtedly, a tactic universally applied to solve many real life problems.

Positive and negative preferences might seem as just two different criteria to reason with, and thus techniques such as those usually adopted by multi-criteria optimization [13] could appear suitable for dealing with them. However, this interpretation would hide the fundamental nature of bipolar preferences, that is, positive preferences are naturally opposite of negative preferences. Moreover, in multi-criteria optimization it is often reasonable to use a Pareto-like approach, thus associating tuples of values to each solution, and comparing solutions according to tuple dominance. Instead, in bipolar problems, it would be very unnatural to force such an approach in all contexts, or to associate to a solution a preference which is neither a positive nor a negative one.

Soft constraints [5] are a useful formalism to model problems with quantitative preferences. However, they can only model negative preferences, since in this framework preference combination returns lower preferences. In this paper we adopt the soft constraint formalism based on semirings to model negative preferences. We then define a new algebraic structure to model positive preferences. To model bipolar problems, we link these two structures and we set the highest negative preference to coincide with the lowest positive preference to model indifference. We then define a combination operator between positive and negative preferences to model preference compensation, and we study its properties.

Non-associativity of preference compensation occurs in many contexts, thus we think it is too restrictive to focus just on associative environments. For example, non-associativity of compensation arises when either positive or negative preferences are aggregated with an idempotent operator (such as min or max), while compensation is instead non-idempotent (such as sum). Our framework allows for non-associativity, since we want to give complete freedom to choose the positive and negative algebraic structures. However, we also describe a technique that, given a negative structure, builds a corresponding positive structure and an associative compensation operator.

Finally, we consider the problem of finding optimal solutions of bipolar problems, by suggesting a possible adaptation of constraint propagation and branch and bound to the generalized scenario.

Summarizing, the main results are:

- a formal definition of an algebraic structure to model bipolar preferences;
- the study of the notion of compensation and of its properties (such as associativity);
- a technique to build a bipolar preference structure with an associative compensation operator;
- the adaptation of branch and bound to solve bipolar problems;
- the definition of bipolar propagation and its use within a branch and bound solver.

The paper is organized as follows. Section 2 recalls the main notions of semiring-based soft constraints. Section 3 describes how to model negative preferences using usual soft constraints, and how to model positive preferences. Section 4 shows how to model both positive and negative preferences, and Section 5 defines constraint problems with both positive and negative preferences. Section 6 shows that it is important to have a bipolar structure for expressing both positive and negative preferences, since expressing all the problems' requirements in a positive (or negative) form might lead to different optimal solutions. Section 7 shows that very often the compensation operator is not associative and it describes a technique to build a bipolar preference structure with an associative compensation operator. Section 8 shows how to adapt branch and bound to solve bipolar problems, how to define bipolar propagation and its use within a branch and bound solver. Finally, Section 9 describes the existing related work and gives some hints for future work.

Earlier versions of parts of this paper have appeared in [6].

2   Semiring-Based Soft Constraints

A soft constraint [5] is a classical constraint [1] where each instantiation of its variables has an associated value from a (totally or partially ordered) set. This set has two operations, which makes it similar to a semiring, and is called a c-semiring. A c-semiring is a tuple $(A, +, \times, 0, 1)$ s.t. A is a set and $0, 1 \in A$; $+$ is commutative, associative, idempotent, 0 is its unit element, and 1 is its absorbing element; $\times$ is associative, commutative, distributes over $+$, 1 is its unit element and 0 is its absorbing element. Given the relation $\leq_S$ over A s.t. $a \leq_S b$ iff $a + b = b$, $\leq_S$ is a partial order; $+$ and $\times$ are monotone on $\leq_S$; 0 is its minimum and 1 its maximum; $\langle A, \leq_S \rangle$ is a lattice and, for all $a, b \in A, a + b = lub(a, b)$. Moreover, if $\times$ is idempotent, then $\langle A, \leq_S \rangle$ is a distributive lattice and $\times$ is its glb. Informally, the relation $\leq_S$ gives us a way to compare (some of the) tuples of values and constraints. In fact, when we have $a \leq_S b$, we will say that b *is better than* a.

Given a c-semiring $S = (A, +, \times, 0, 1)$, a finite set D (the domain of the variables), and an ordered set of variables V, a constraint is a pair $\langle def, con \rangle$ where $con \subseteq V$ and $def : D^{|con|} \rightarrow A$. Therefore, a constraint specifies a set of variables (the ones in con), and assigns to each tuple of values of D of these variables an element of the semiring set A. Given a subset of variables $I \subseteq V$, and a soft constraint $c = \langle def, con \rangle$, the projection of c over I, written $c \Downarrow_I$, is a new soft constraint $\langle def', con' \rangle$, where $con' = con \cap I$ and $def'(t') = \sum_{\{t | t \downarrow_{con'}^{con} = t'\}} def(t)$. The scope, con', of the projection constraint contains the variables that con and I have in common, and thus $con' \subseteq con$. Moreover, the preference associated to each assignment to the variables in con', denoted with t', is the highest ($\sum$ is the additive operator of the c-semiring) among the preferences associated by def to any completion of t', t, to an assignment to con.

A soft constraint satisfaction problem (SCSP) is just a set of soft constraints over a set of variables. A classical CSP is just an SCSP where the chosen c-semiring is: $S_{CSP} = (\{false, true\}, \vee, \wedge, false, true)$. On the other hand, fuzzy CSPs can be modelled in the SCSP framework by choosing the c-semiring: $S_{FCSP} = ([0, 1], max, min, 0, 1)$. For weighted CSPs, the semiring is $S_{WCSP} = (R^+, min, +, +\infty, 0)$. Here preferences are interpreted as costs of which we want to minimize the sum. For probabilistic CSPs, the semiring is $S_{PCSP} = ([0, 1], max, \times, 0, 1)$. Here preferences are interpreted as probabilities and the aim is to maximize the joint probability.

Given an assignment to all the variables of an SCSP, we can compute its preference value by combining the preferences associated by each constraint to the subtuples of the assignments referring to the variables of the constraint. An optimal solution of an

SCSP is then a complete assignment t such that there is no other complete assignment t' with $pref(t) <_S pref(t')$.

3 Negative and Positive Preferences

The structure we use to model negative preferences is exactly a c-semiring, as defined in Section 2. In fact, in a c-semiring the element which acts as indifference is 1, since $\forall a \in A, a \times 1 = a$. This element is the best in the ordering, which is consistent with the fact that indifference is the best preference when using only negative preferences. Moreover, in a c-semiring, combination goes down in the ordering, since $a \times b \leq a, b$. This can be naturally interpreted as the fact that combining negative preferences worsens the overall preference. From now on, we use $(N, +_n, \times_n, \perp_n, \top_n)$ as c-semiring to model negative preferences.

When dealing with positive preferences, we want two main properties to hold: combination should bring to better preferences, and indifference should be lower than all the other positive preferences. These properties can be found in the following structure.

Definition 1. *A positive preference structure is a tuple* $(P, +_p, \times_p, \perp_p, \top_p)$ *s.t.*

- *P is a set and $\top_p, \perp_p \in P$;*
- *$+_p$, the additive operator, is commutative, associative, idempotent, with $\perp_p$ as its unit element ($\forall a \in P, a +_p \perp_p = a$) and $\top_p$ as its absorbing element ($\forall a \in P, a +_p \top_p = \top_p$);*
- *$\times_p$, the multiplicative operator, is associative, commutative and distributes over $+_p$ ($a \times_p (b +_p c) = (a \times_p b) +_p (a \times_p c)$), with $\perp_p$ as its unit element and $\top_p$ as its absorbing element[1].*

Notice that the additive operator of this structure has the same properties as the corresponding one in c-semirings, and thus it induces a partial order over P in the usual way: $a \leq_p b$ iff $a +_p b = b$. This allows to prove that $+_p$ is monotone over $\leq_p$ and that it is the least upper bound in the lattice $(P, \leq_p)$. On the other hand, the multiplicative operator has different properties. More precisely, the best element in the ordering ($\top_p$) is now its absorbing element, and the worst element ($\perp_p$) is its unit element. This reflects the desired behavior of the combination of positive preferences.

Theorem 1. *Given a positive preference structure* $(P, +_p, \times_p, \perp_p, \top_p)$, *consider the relation* $\leq_p$ *over P. Then,* $\times_p$ *is monotone over* $\leq_p$ *(that is, for any $a, b \in P$ s.t. $a \leq_p b$, then $a \times_p d \leq_p b \times_p d, \forall d \in P$), and $\forall a, b \in P$, $a \times_p b \geq_p a +_p b \geq_p a, b$.*

Proof. Since $a \leq_p b$ iff $a +_p b = b$, then $b \times_p d = (a +_p b) \times_p d = (a \times_p d) +_p (b \times_p d)$. Thus $a \times_p d \leq_p b \times_p d$. Also, $a \times_p b = a \times_p (b +_p \perp_p) = (a \times_p b) +_p (a \times_p \perp_p) = (a \times_p b) +_p a$. Thus $a \times_p b \geq_p a$ (the same for b). Finally: $a \times_p b \geq a, b$. Thus $a \times_p b \geq lub(a, b) = a +_p b$.   Q.E.D.

In a positive preference structure, $\perp_p$ is the element modelling indifference. In fact, it is the worst one in the ordering and it is the unit element for the combination operator

---

[1] The absorbing nature of $\top_p$ can be derived from the other properties.

$\times_p$. These are exactly the desired properties for indifference w.r.t. positive preferences. The role of $\top_p$ is to model a very high preference, much higher than all the others. In fact, since it is the absorbing element of the combination operator, when we combine any positive preference a with $\top_p$, we get $\top_p$.

An example of a positive preference structure is $P_1 = (R^+, max, +, 0, +\infty)$, where preferences are positive reals. The smallest preference that can be assigned is 0. It represents the lack of any positive aspect and can thus be regarded as indifference. Preferences are aggregated taking the sum and are compared taking the max.

Another example is $P_2 = ([0, 1], max, max, 0, 1)$. In this case preferences are reals between 0 and 1, as in the fuzzy semiring for negative preferences. However, the combination operator is max, which gives, as a resulting preference, the highest one among all those combined.

As an example of a partially ordered positive preference structure consider the Cartesian product of the two structures described above: $(R^+\times[0, 1], (max, max), (+, max), (0, 0), (+\infty, 1))$. Positive preferences, here, are ordered pairs where the first element is a positive preference of type P_1 and the second one is a positive preference of type P_2. Consider for example the (incomparable) pairs $(8, 0.1)$ and $(3, 0.8)$. Applying the aggregation operator (+,max) gives the pair $(11, 0.8)$ which, as expected, is better than both pairs, since $max(8, 3, 11) = 11$ and $max(0.1, 0.8, 0.8) = 0.8$.

4   Bipolar Preference Structures

Once we are given a positive and a negative preference structure, a naive way to combine them would be performing the Cartesian product of the two structures. For example, if we have a positive structure P and a negative structure N, taking their Cartesian product would mean that, given a solution, it will be associated with a pair (p, n), where $p \in P$ is the overall positive preference and $n \in N$ is the overall negative preference. Such a pair is in general neither an element of P nor of N, so it is neither positive nor negative, unless one or both of p and n are the indifference element. Moreover, the ordering induced over these pairs is the well known Pareto ordering, which may induce a lot of incomparability among the solutions. These two features imply that compensation is not allowed at all. Instead, we believe that it should be allowed, if desired. We will therefore now describe a bipolar preference structure that allows for it.

Definition 2. *A bipolar preference structure is a tuple* $(N, P, +, \times, \bot, \Box, \top)$, *where*

- $(P, +_{|P}, \times_{|P}, \Box, \top)$ *is a positive preference structure;*
- $(N, +_{|N}, \times_{|N}, \bot, \Box)$ *is a c-semiring;*
- $+ : (N \cup P)^2 \longrightarrow (N \cup P)$ *is s.t.* $a_n + a_p = a_p$ *for any* $a_n \in N$ *and* $a_p \in P$; *this operator induces as partial ordering on* $N \cup P$: $\forall a, b \in P \cup N$, $a \leq b$ *iff* $a+b = b$;
- $\times : (N \cup P)^2 \longrightarrow (N \cup P)$ *is an operator (called the* compensation *operator) that, for all* $a, b, c \in N \cup P$, *satisfies the following properties:*
  - *commutativity:* $a \times b = b \times a$;
  - *monotonicity: if* $a \leq b$, *then* $a \times c \leq b \times c$.

In the following, we will write $+_n$ instead of $+_{|_N}$ and $+_p$ instead of $+_{|_P}$. Similarly for $\times_n$ and $\times_p$. Moreover, we will sometimes write $\times_{np}$ when operator $\times$ will be applied to a pair in $(N \times P)$.

Notice that bipolar structures generalize both negative and positive preference structures via a bipolar structure with a single positive/negative preference element.

Given the way the ordering is induced by $+$ on $N \cup P$, easily, we have $\bot \leq \square \leq \top$. Thus, there is a unique maximum element (that is, $\top$), a unique minimum element (that is, $\bot$); the element $\square$ is smaller than any positive preference and greater than any negative preference, and it is used to model indifference. The shape of a bipolar preference structure is shown in the following figure:

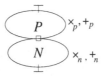

Notice that, although all positive preferences are strictly above negative preferences, our framework does not prevent from using the same scale, or partially overlapping scales, to represent positive and negative preferences.

A bipolar preference structure allows us to have different ways to model and reason about positive and negative preferences. In fact, we can have different lattices $(P, \leq_p)$ and $(N, \leq_n)$. This is common in real-life problems, where negative and positive statements are not necessarily expressed using the same granularity. For example, we could be satisfied with just two levels of negative preferences, while requiring several levels of positive preferences. Nevertheless, our framework allows to model cases in which the two structures are isomorphic, as well (see Section 7).

Notice that the combination of a positive and a negative preference is a preference which is higher than, or equal to, the negative one and lower than, or equal to, the positive one.

Theorem 2. *Given a bipolar preference structure* $(N, P, +, \times, \bot, \square, \top)$*, for all* $p \in P$ *and* $n \in N$*,* $n \leq p \times n \leq p$*.*

Proof. For any $n \in N$ and $p \in P$, $\square \leq p$ and $n \leq \square$. By monotonicity of $\times$, we have: $n \times \square \leq n \times p$ and $n \times p \leq \square \times p$. Hence: $n = n \times \square \leq n \times p \leq \square \times p = p$. Q.E.D.

This means that the compensation of positive and negative preferences lies in one of the chains between the two combined preferences. Notice that all such chains pass through the indifference element $\square$. Possible choices for combining strictly positive with strictly negative preferences are thus the average or the median operator, or also the minimum or the maximum.

Moreover, by monotonicity, we can show that if $\top \times \bot = \bot$, then the result of the compensation between any positive preference and the bottom element is the bottom element, and if $\top \times \bot = \top$, then the compensation between any negative preference and the top element is the top element.

Theorem 3. *Given a bipolar preference structure* $(N, P, +, \times, \bot, \square, \top)$*, if* $\top \times \bot = \bot$*, then* $\forall p \in P$*,* $p \times \bot = \bot$*, while, if* $\top \times \bot = \top$*, then* $\forall n \in N$*,* $n \times \top = \top$*.*

Proof. Assume $\top \times \bot = \bot$. Since for all $p \in P, p \leq \top$, then, by monotonicity of $\times$, $p \times \bot \leq \top \times \bot = \bot$, hence $p \times \bot = \bot$.

Assume $\top \times \bot = \top$. Since for all $n \in N, \bot \leq n$, then, by monotonicity of $\times$, $\top = \top \times \bot \leq \top \times n$, hence $\top \times n = \top$.   Q.E.D

In the following table we give three examples of bipolar preference structures, one for each row.

N,P	$+$	$\times$	$\bot, \square, \top$
R^-, R^+	$+_p=+_n=+_{np}=$ max	$\times_p=\times_n=\times_{np}=$sum	$-\infty, 0, +\infty$
$[-1,0], [0,1]$	$+_p=+_n=+_{np}=$max	$\times_p=$max, $\times_n=$min, $\times_{np}=$sum	$-1, 0, 1$
$[0,1], [1,+\infty]$	$+_p=+_n=+_{np}=$max	$\times_p=\times_n=\times_{np}=$prod	$0, 1, +\infty$

The structure described in the first row uses real numbers as positive and negative preferences. Compensation is obtained by summing the preferences, while the ordering is given by the max operator. In the second structure we have positive preferences between 0 and 1 and negative preferences between -1 and 0. Aggregation of positive preferences is max and of negative preferences is min, while compensation between positive and negative preferences is sum, and the order is given by max. In the third structure we use positive preferences between 1 and $+\infty$ and negative preferences between 0 and 1. Compensation is obtained by multiplying the preferences and ordering is again via max. Notice that if $\top \times \bot \in \{\top, \bot\}$, then compensation in the first and in the third structure is associative.

5 Bipolar Preference Problems

A bipolar constraint is just a constraint where each assignment of values to its variables is associated to one of the elements in a bipolar preference structure. A bipolar CSP (V, C) is then just a set of variables V and a set of bipolar constraints C over V. There could be many ways of defining the optimal solutions of a bipolar CSP. Here we propose one which avoids problems due to the possible non-associativity of the compensation operator, since compensation never involves more than two preference values.

Definition 3. *Given a bipolar preference structure* $(N, P, +, \times, \bot, \square, \top)$, *a solution of a bipolar CSP* (V, C) *is a complete assignment to all variables in* V, *say* s, *and an associated preference which is computed as follows:* $pref(s) = (p_1 \times_p \ldots \times_p p_k) \times (n_1 \times_n \ldots \times_n n_l)$, *where* $p_i \in P$ *for* $i := 1, \ldots, k$ *and* $n_j \in N$ *for* $j := 1, \ldots, l$ *and where* $\exists \langle def_i, con_i \rangle \in C$ *s.t.* $p_i = def_i(s \downarrow_{con})$, *and* $\exists \langle def_j, con_j \rangle \in C$ *s.t.* $n_j = def_j(s \downarrow_{con})$. *A solution* s *is an optimal solution if there is no other solution* s' *with* $pref(s') > pref(s)$.

In this definition, the preference of a solution s is obtained by combining all the positive preferences associated to its projections over the constraints, by using $\times_p$, combining all the negative preferences associated to its projections over the constraints, by using $\times_n$, and then, combining the two preferences obtained so far (one positive and one negative) by using the operator $\times_{np}$.

Such a definition follows the same idea proposed in Chapter IV of [4] for evaluating the tendency of an act. Such an idea consists of summing up all the values of all the pleasures produced by the considered act on one side, and those of all the pains produced by it on the other, and then balancing these two resulting values in a value which can be on the side of pleasure or on the side of pain. If this value is on the side of pleasure, then the tendency of the act is good, otherwise the tendency is bad.

Consider the scenario in which we want to buy a car and we have preferences over some features. In terms of color, we like red, we are indifferent to white, and we hate black. Also, we like convertible cars a lot and we don't care much for big cars (e.g., SUVs). In terms of engines, we like diesel. However, we don't want a diesel convertible.

We represent positive preferences via positive integers, negative preferences via negative integers and we maximize the sum of all kinds of preferences. This can be modelled by a bipolar preference structure where $N = [-\infty, 0]$, $P = [0, +\infty]$, $+ =$ max, $\times =$ sum, $\perp = -\infty$, $\square = 0$, $\top = +\infty$.

The following figure shows the structure (variables, domains, constraints, and preferences) of such a bipolar CSP, where preferences have been chosen to fit the informal specification above, and 0 is used to model indifference (also when tuples are not shown).

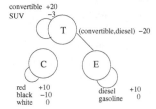

Consider solution $s_1 =$ (red,convertible,diesel). We have $pref(s_1) = (def_1(\text{red}) \times def_2(\text{convertible}) \times def_3(\text{diesel})) \times def_4(\text{convertible, diesel}) = (10 + 20 + 10) + (-20) = 20$. We can see that the optimal solution is (red, convertible, gasoline) with global preference of 30.

Consider now a different bipolar preference structure, which differs from the previous one only for $\times_p$, which is now max. Now solution s_1 has preference $pref(s_1) = (def_1(\text{red}) \times def_2(\text{convertible}) \times def_3(\text{diesel})) \times def_4(\text{convertible, diesel}) = max(10, 20, 10) + (-20) = 0$. It is easy to see that now an optimal solution has preference 20. There are two of such solutions: one is the same as the optimal solution above, and the other one is (white,convertible, gasoline). The two cars have the same features except for the color. A white convertible is just as good as a red convertible because we decided to aggregate positive preference by taking the maximum elements rather than by summing them.

6   Positive Versus Negative Preferences

Positive and negative preferences look so similar that, even though we know they need different combination operators, we could wonder why we need two different structures to handle them. Why can't we just have one structure, for example the negative one, and transform each positive preference into a negative one? For example, if there are only two colors for cars, (i.e., red and blue), and we only like blue, instead of saying this using positive preferences (i.e., we like blue with a certain positive preference), we could phrase it using negative preferences (i.e., we don't like red with a certain negative preference). In other words, instead of associating a positive preference to blue and indifference to red, we could give a negative preference to red and indifference to blue.

In this section we will show that, by doing this, we could modify the solution ordering, thus representing a different optimization problem. Thus sometimes the two preference structures are needed to model the problems under consideration: using just one of them would not suffice.

It is easy to show that, by moving from a positive to a negative modelling of the same information, as we have done in the example above, all solutions get a lower preference value. In fact, in this transformation, a positive preference is replaced by indifference, or indifference is replaced by a negative preference. So, in any case, some preference is replaced by a lower one, and by monotonicity of the aggregation operators ($\times_n$, $\times_p$, and $\times$), the overall preference of the solutions is lower as well.

However, it is worth noting that this preference lowering might not preserve the ordering among solutions. That is, solutions that were ordered in a certain way before the modification, can be ordered in the opposite way after it. This is due to the fact that aggregation of positive and negative preferences may behave differently. The following example shows this phenomenon.

Consider the bipolar preference structure $(R^-, R^+, max, \times, -\infty, 0, +\infty)$, where $\times$ is such that $\times_p = \times_{np} = sum$ and $\times_n = min$. This means that we want to maximize the sum of positive preferences, maximize the minimal negative preference (thus negative preferences are handled as fuzzy constraints), that positive preferences are between 0 and $+\infty$, and negative preferences are between 0 and $-\infty$. Compensation is via algebraic sum, thus values v and $-v$ are compensated completely (that is, the result of the compensation is 0), while the compensation of values v and $-v'$ is $v - v'$.

Consider now a bipolar CSP over this structure with four variables, say X, Y, Z, W, where each variable has a Boolean domain as follows: $D(X) = \{a, \bar{a}\}$, $D(Y) = \{b, \bar{b}\}$, $D(Z) = \{c, \bar{c}\}$, and $D(W) = \{d, \bar{d}\}$. Assume now that the preference of a is 2, of b is 1, of c is 2.4, and of d is 0.5, while the preference of the other elements is indifference (that is, 0 in this example). This means that we have expressed all our statements in a positive form.

Consider now two solutions s and s' as follows: $s = (a, b, \bar{c}, \bar{d})$ and $s' = (\bar{a}, \bar{b}, c, d)$. By computing the preference of s, we get $(2 + 1) + min(0, 0) = 3$, while for s' we get $min(0, 0) + (2.4 + 0.5) = 2.9$. Thus s is better than s'.

Assume now to express the same statements in negative terms assuming that if we like at level p an assignment t, then we dislike $\bar{t}$ at the same level p. Hence, the preference of $\bar{a}$ is -2, of $\bar{b}$ is -1, of $\bar{c}$ is -2.4, and of $\bar{d}$ is -0.5, while the preference of the other elements is 0. Now the preference of s is $(0 + 0) + min(-2.4, -0.5) = -2.4$, while the preference of s' is $min(-2, -1) + (0 + 0) = -2$. Thus s' is better than s.

7   Associativity of Preference Compensation

In general, the compensation operator $\times$ may be not associative. Here we list some sufficient conditions for the non-associativity of the $\times$ operator.

Theorem 4. *Given a bipolar preference structure* $(P, N, +, \times, \bot, \square, \top)$, *operator* $\times$ *is not associative if at least one of the following two conditions is satisfied:*

- $\top \times \bot = c \in (N \cup P) - \{\top, \bot\}$;
- $\exists p \in P - \{\top, \square\}$ *and* $n \in N - \{\bot, \square\}$ *s.t.* $p \times n = \square$ *and at least one of the following conditions holds:*
    - $\times_p$ *or* $\times_n$ *is idempotent;*
    - $\exists p' \in P - \{p, \top\}$ *s.t.* $p' \times n = \square$ *or* $\exists n' \in N - \{n, \bot\}$ *s.t.* $p \times n' = \square$;
    - $\top \times \bot = \bot$ *and* $\exists n' \in N - \{\bot\}$ *s.t.* $n \times n' = \bot$;
    - $\top \times \bot = \top$ *and* $\exists p' \in P - \{\top\}$ *s.t.* $p \times p' = \top$;
    - $\exists a, c \in N \cup P$ *s.t.* $a \times p = c$ *iff* $c \times n \neq a$ *(or* $\exists a, c \in N \cup P$ *s.t.* $a \times n = c$ *iff* $c \times p \neq a$).

Proof. If $c \in P - \{\top\}$, then $\top \times (\top \times \bot) = \top \times c = \top$, while $(\top \times \top) \times \bot = \top \times \bot = c$. If $c \in N - \{\bot\}$, then $\bot \times (\bot \times \top) = \bot \times c = \bot$, while $(\bot \times \bot) \times \top = \bot \times \top = c$.

Assume that $\exists p \in P - \{\top, \square\}$ and $n \in N - \{\bot, \square\}$ s.t. $p \times n = \square$. If $\times_p$ is idempotent, then $p \times (p \times n) = p \times \square = p$, while $(p \times p) \times n = p \times n = \square$. Similarly if $\times_n$ is idempotent.

If $\exists p' \in P - \{p, \top\}$ s.t. $p' \times n = \square$, then $(p \times n) \times p' = p'$, while $p \times (n \times p') = p$. Analogously, if $\exists n' \in N - \{n, \bot\}$ s.t. $p \times n' = \square$.

If $\top \times \bot = \bot$, then, by Theorem 3, $p \times \bot = \bot$. If $\exists n' \in N - \{\bot\}$ s.t. $n \times n' = \bot$, then $(p \times n) \times n' = \square \times n' = n'$, while $p \times (n \times n') = p \times \bot = \bot \neq n'$.

If $\top \times \bot = \top$, then, by Theorem 3, $n \times \top = \top$. If $\exists p' \in P - \{\top\}$ s.t. $p \times p' = \top$, then $(n \times p) \times p' = \square \times p' = p'$, while $n \times (p \times p') = n \times \top = \top \neq p'$.

If $c \times n \neq a$, then $(a \times p) \times n = c \times n \neq a$, but $a \times (p \times n) = a \times \square = a$. Analogously if $c \times p \neq a$.　　　　　　　　　　　　　　　　　　　　Q.E.D.

Notice that sufficient conditions refer to various aspects of a bipolar preference structure: properties of operators, shape of P and N orderings, the relation between $\times$ and the other operators. Since some of these conditions often occur in practice, it is not reasonable to require always associativity of $\times$.

It is however useful to be able to build bipolar preference structures where compensation is associative. It is obvious that, if we are free to choose any positive and any negative preference structure when building the bipolar framework, we will never be able to assure associativity of the compensation operator. Thus, to assure this, we must pose some restrictions on the way a bipolar preference structure is built.

We describe now a procedure to build positive preferences as inverses of negative preferences, that assures that the resulting bipolar preference structure has an associative compensation operator. To do that, $\times_n$ must be non-idempotent. The methodology is called *localization* and represents a standard systematic technique for adding multiplicative inverses to a (semi)ring [8].

Given a (semi)ring with carrier set N (representing, in our context, a negative preference structure), and a subset $S \subseteq N$, we can construct another structure with carrier set P (representing, for us, a positive preference structure), and a mapping from N to P which makes all elements in the image of S invertible in P. The localization of N by S is also denoted by $S^{-1}N$.

We can select any subset S of N. However, it is usual to select a subset S of N which is closed under $\times_n$, such that $1 \in S$ (1 is the unit for $\times_n$, which represents indifference), and $0 \notin S$.

Given N and S, let us consider the quotient field of N w.r.t. S. This is denoted by $Quot(N, S)$, and will represent the carrier set of our bipolar structure. One can construct $Quot(N, S)$ by just taking the set of equivalence classes of pairs (n, d), where n and d are elements of N and S respectively, and the equivalence relation is: $(n, d) \equiv (m, b) \iff n \times_n b = m \times_n d$. We can think of the class of (n, d) as the fraction $\frac{n}{d}$.

The embedding of N in $Quot(N, S)$ is given by the mapping $f(n) = (n, 1)$, thus the (semi)ring N is a sub(semi)ring of $S^{-1}N$ via the identification $f(a) = \frac{a}{1}$.

The next step is to define the $+$ and $\times$ operator in $Quot(N, S)$, as function of the operators $+_n$ and $\times_n$ of N. We define $(n, d) + (m, b) = ((n \times_n b) +_n (m \times_n d), d \times_n b)$ and $(n, d) \times (m, b) = (m \times_n n, d \times_n b)$. By using the fraction representation we obtain the usual form where the addition and the multiplication of the formal fractions are defined according to the natural rules: $\frac{a}{s} + \frac{b}{t} = \frac{(a \times_n t) +_n (b \times_n s)}{s \times_n t}$ and $\frac{a}{s} \times \frac{b}{t} = \frac{a \times_n b}{s \times_n t}$.

It can be shown that the structure $(P, +_p, \times_p, \frac{1}{1}, \frac{1}{0})$, where $P = \{\frac{1}{q}$ s.t. $a \in (S \cup \{0\})\}$, $+_p$ and $\times_p$ are the operators $+$ and $\times$ restricted over $\frac{1}{S} \times \frac{1}{S}$, $\frac{1}{1}$ is the bottom element in the induced order (notice that the element coincide with 1), and $\frac{1}{0}$ is the top element of the structure[2], is a positive preference structure. Moreover, $Quot(N, S) = P \cup N$, and it is the carrier of a bipolar preference structure $\langle P, N, +, \times, 0, \frac{1}{1}, \frac{1}{0} \rangle$ where $\times$ is an associative compensation operator by construction.

Notice that the first example of the table in Section 4, as well as the third example restricted to rational numbers, can be obtained via the localization procedure.

8   Solving Bipolar Preference Problems

Bipolar problems are NP-complete, since they generalize both classical and soft constraints, which are already known to be difficult problems [5]. In this section we will consider how to adapt some usual techniques for soft constraints to bipolar problems.

8.1   Branch and Bound

Preference problems based on c-semirings can be solved via a branch and bound technique (BB), possibly augmented via soft constraint propagation, which may lower the preferences and thus allow for the computation of better bounds [5].

In bipolar CSPs, we have both positive and negative preferences. However, if the compensation operator is associative, standard BB can be applied. Thus bipolar preferences can be handled without additional effort.

However, if compensation is not associative, the upper bound computation has to be slightly changed to avoid performing compensation before all the positive and the negative preferences have been collected. More precisely, each node of the search tree is associated to a positive and a negative preference, say p and n, which are obtained by aggregating all preferences of the same type obtained in the instantiated part of the problem. Next, all the best preferences (which may be positive or negative) in the

---

[2] This element is introduced ad hoc because 0 is not an unit and cannot be used to build its inverse.

uninstantiated part of the problem are considered. By aggregating those of the same type, we get a positive and a negative preference, say p' and n', which can be combined with the ones associated to the current node. This produces the upper bound $ub = (p \times_p p') \times (n \times_n n')$, where $p' = p_1 \times_p \ldots \times_p p_w$, $n' = n_1 \times_n \ldots \times_n n_s$, with $w + s = r$, where r is the number of uninstantiated variables/constraints. Notice that ub is computed via $r - 1$ aggregation steps and one compensation step.

On the other hand, when compensation is associative, we don't need to postpone compensation until all constraints have been considered. Thus, ub can be computed as $ub = a_1 \times \ldots \times a_{p+r}$, where $a_i \in N \cup P$ is the best preference found in a constraint of either the instantiated part of the problem (first p elements) or the uninstantiated part of the problem (last r elements). Thus ub can be computed via at most $p + r - 1$ steps among which there can be many compensation steps.

8.2 Bipolar Propagation

When looking for an optimal solution, BB can be helped by some form of partial or full constraint propagation. To see whether this can be done when solving bipolar problems as well, we must first understand what constraint propagation means in such problems. For sake of semplicity, we will focus here on arc-consistency.

Given any bipolar constraint, let us first define its negative version $neg(c)$, which is obtained by just replacing the positive preferences with indifference. Similarly, the positive version $pos(c)$ is obtained by replacing negative preferences with indifference.

A binary bipolar constraint c is then said negatively arc-consistent (NAC) iff $neg(c)$ is soft arc-consistent. If the binary constraint connects variables X and Y, let us call it c_{XY}, and let us call c_X the soft domain of X and c_Y the soft domain of Y. Then, being soft arc consistent means that $neg(c_X) = (neg(c_X) \times_n neg(c_Y) \times_n neg(c_{XY})) \Downarrow_X$ and $neg(c_Y) = (neg(c_X) \times_n neg(c_Y) \times_n neg(c_{XY})) \Downarrow_Y$. If this is not so, we can make a binary bipolar constraint NAC by modifying the soft domains of its two variables such that the two equations above hold. The modifications required can only decrease some preference values. Thus some negative preferences can become more negative than before. If operator $\times_n$ is idempotent, then such modifications generate a new constraint which is equivalent to the given one.

Let us now consider the positive version of a constraint. Let us also define an operation $\Uparrow_X$, which, taken any constraint c_S over variables S such that $X \in S$, computes a new constraint over X as follows: for every value a in the domain of X, its preference is computed by taking the greatest lower bound of all preferences given by c_S to tuples containing $X = a$. Then we say that a binary bipolar constraint is positively arc-consistent (PAC) iff $c_X = (pos(c_X) \times_p pos(c_Y) \times_p pos(c_{XY})) \Uparrow_X$ and $c_Y = (pos(c_X) \times_p pos(c_Y) \times_p pos(c_{XY})) \Uparrow_Y$. If this is not so, we can make a binary bipolar constraint PAC by modifying the soft domains of its two variables such that the two equations above hold. The modifications required can only involve the increase of some preference values. Thus some positive preferences can become more positive than before. If operator $\times_p$ is idempotent, such modifications generate a new constraint which is equivalent to the given one.

Finally, we say that a binary bipolar constraint is Bipolar Arc-Consistent (BAC) iff it is NAC and PAC. A bipolar constraint problem is BAC iff all its constraints are BAC.

If a bipolar constraint problem is not BAC, we can consider its negative and positive versions and achieve PAC and NAC on them. If both $\times_n$ and $\times_p$ are idempotent, this can be seen as the application of functions which are monotone, inflationary, and idempotent on a suitable partial order. Thus usual algorithms based on chaotic iterations [1] can be used, with the assurance of terminating and having a unique equivalent result which is independent of the order in which constraints are considered. However, this can generate two versions of the problem (of which one is NAC and the other one is PAC) which could be impossible to reconcile into a single bipolar problem.

The problem can be solved by achieving only partial forms of PAC and NAC in a bipolar problem. The basic idea is to consider the given bipolar problem, apply the NAC and PAC algorithms to its negative and positive versions, and then modify the preferences of the original problem only when the two new versions can be reconciled, that is, when at least one of the two new preferences is the indifference element. In fact, this means that, in one of the two consistency algorithms, no change has been made. If this holds, the other preference is used to modify the original one. This algorithm achieves a partial form of BAC, that we call p-BAC, and assures equivalence.

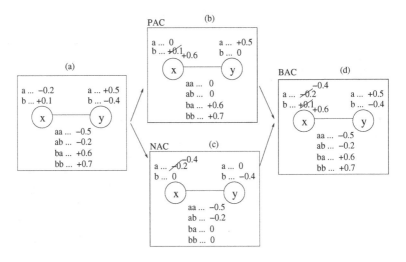

Fig. 1. How to make a bipolar constraint p-BAC

In Figure 1 it is shown how to make a bipolar constraint partially Bipolar Arc-Consistent. Part (a) shows a bipolar constraint, named c_X, over variable X, a bipolar constraint, named c_Y, over variable Y, and a bipolar constraint, named c_{XY}, linking X and Y. Preferences are modelled by the bipolar preference structure ($N = [-1, 0]$, $P = [0, 1]$, $+ = max$, $\times$, $\bot = -1$, $\square = 0$, $\top = 1$), where $\times$ is such that $\times_p = max$, $\times_n = min$ and $\times_{np} = sum$. Since preferences are given independently in c_X, c_Y and c_{XY}, it is possible to give a low positive preference for a value of X (e.g., $X = b$) in c_X, a negative preference for a value of Y (e.g., $X = b$) in C_Y, but an high positive preference for the combination of such values in c_{XY}. In Part (b) we present the positive version of c_{XY}, that becomes PAC, by increasing the positive preference associated to

$X = b$ from $+0.1$ to $+0.6$. Part (c) presents the negative version of c_{XY}, that becomes NAC, by decreasing the negative preference associated to $X = a$ from -0.2 to -0.4. In Part (d) we show how to achieve p-BAC of c_{XY}. For obtaining p-BAC we must reconcile the modified preferences obtained in Part (b) and in Part (c) when it is possible. Since in this example it is always possible to reconcile such preferences, we obtain a bipolar constraint which is not only p-BAC, but also BAC.

In this approach we require idempotency of $\times_p$ and $\times_n$. However, we could apply arc-consistency also when such operators are not idempotent, by following the extended version of arc-consistency presented in [10,7,18].

Notice that our algorithm will possibly decrease some negative preferences and increase some positive preferences. Therefore, if we use constraint propagation to improve the bounds in a BB algorithm, it will actually sometimes produce worse bounds, due to the increase of the positive preferences. We will thus use only the propagation of negative preferences (that is, NAC) within a BB algorithm. Since the upper bound is just a combination of several preferences, and since preference combination is monotonic, lower preferences give a lower, and thus better, upper bound.

9   Related and Future Work

Bipolarity is an important topic in several fields, such as psychology [9,16,19,20] and multi-criteria decision making [15] and it has recently attracted interest in the AI community, especially in argumentation [17] and qualitative reasoning [2,3,11,12]. These works consider how two alternatives should be compared, given for each a set of positive arguments and a set of negative ones, but they don't analyze the question of combinatorial choice.

Two works in qualitative reasoning which are directly related to our approach are those described in [2,3]. In such papers a bipolar preference model based on a fuzzy-possibilistic approach is described where fuzzy preferences are considered and negative preferences are interpreted as violations of constraints. In this case precedence is given to negative preference optimization, and positive preferences are only used to distinguish among the optimals found in the first phase, thus not allowing for compensation.

Another related work is [14], which considers only totally ordered unipolar and bipolar preference scales, but not partially ordered bipolar scales like us. When the preferences are totally ordered, operators $\times_n$ and $\times_p$ described here correspond respectively to the $t-norm$ and $t-conorm$ used in [14]. Moreover, in [14] it is defined an operator, the $uninorm$, which can be seen as a restricted form of compensation and it is forced to always be associative.

We plan to develop a solver for bipolar CSPs, which should be flexible enough to accommodate for both associative and non-associative compensation operators. The outlined algorithms for BB, NAC, PAC, and p-BAC will also be implemented and tested over classes of bipolar problems.

We also intend to consider the presence of uncertainty in bipolar problems, possibly using possibility theory and to develop solving techniques for such scenarios. Another line of future research is the generalization of other preference formalisms, such as multicriteria methods and CP-nets, to deal with bipolar preferences and to study the relation between bipolarization and tradeoffs.

Acknowledgements

This work has been supported by Italian MIUR PRIN project "Constraints and Preferences" (n. 2005015491).

References

1. Apt, K.R.: Principles of Constraint Programming. Cambridge University Press, Cambridge (2003)
2. Benferhat, S., Dubois, D., Kaci, S., Prade, H.: Bipolar representation and fusion of preferences in the possibilistic logic framework. In: KR 2002, Morgan Kaufmann, San Francisco (2002)
3. Benferhat, S., Dubois, D., Kaci, S., Prade, H.: Bipolar possibility theory in prefence modeling: representation, fusion and optimal solutions. Information Fusion, an Int. Journal on Multi-Sensor, Multi-Source Information Fusion (2006)
4. Bentham, J.: An Introduction to the Principles of Morals and Legislation. Oxford University Press, Oxford, 1789
5. Bistarelli, S., Montanari, U., Rossi, F.: Semiring-based constraint solving and optimization. Journal of the ACM 44(2), 201–236 (1997)
6. Bistarelli, S., Pini, M.S., Rossi, F., Venable, K.B.: Bipolar preference problems. In: ECAI-06 (poster), IOS Press, Amsterdam (2006)
7. Bistarelli, S., Gadducci, F.: Enhancing constraints manipulation in semiring-based formalisms. In: ECAI, pp. 63–67. IOS Press, Amsterdam (2006)
8. Bruns, W., Herzog, J.C.-M.: Rings. Cambridge University Press, Cambridge (1998)
9. Suci, G.J., Osgood, C.E., Tannenbaum, P.H.: The Measurement of Meaning. University of Illinois Press, Chicago (1957)
10. Cooper, M., Schiex, T.: Arc consistency for soft constraints. AI Journal 154(1-2), 199–227 (2004)
11. Dubois, D., Fargier, H.: On the qualitative comparison of sets of positive and negative affects. In: Godo, L. (ed.) ECSQARU 2005. LNCS (LNAI), vol. 3571, pp. 305–316. Springer, Heidelberg (2005)
12. Dubois, D., Fargier, H.: Qualitative decision making with bipolar information. In: KR'06, pp. 175–186 (2006)
13. Ehrgott, M., Gandibleux, X.: Multiple Criteria Optimization. State of the art annotated bibliographic surveys. Kluwer Academic Publishers, Dordrecht (2002)
14. Grabisch, M., de Baets, B., Fodor, J.: The quest for rings on bipolar scales. Int. Journ. of Uncertainty, Fuzziness and Knowledge-Based Systems (2003)
15. Grabisch, M., Labreuche, C.: Bi-capacities - parts i and ii. Fuzzy Sets and Systems 151, 211–260 (2005)
16. Gardner, W.L., Cacioppo, J.T., Berntson, G.G.: Beyond bipolar conceptualizations and measures: the case of attitudes and evaluative space. Personality and Social Psychology Review 1, 3–25 (1997)
17. Bonnenfon, J.F., Amgoud, L., Prade, H.: An argumentation-based approach to multiple criteria decision. In: Godo, L. (ed.) ECSQARU 2005. LNCS (LNAI), vol. 3571, pp. 10–17. Springer, Heidelberg (2005)
18. Larrosa, J., Schiex, T.: In the quest of the best form of local consistency for weighted csp. In: Proc. 18th IJCAI, pp. 239–244. Morgan Kaufman, San Francisco (2003)
19. Peters, E., Slovic, P., Finucane, M., MagGregor, D.G.: Rational actors or rational fools? Implications of the affect heuristic for behavioural economics. The Journal of Socio-Economics 31, 329–342 (2002)
20. Tversky, A., Kahneman, D.: Advances in prospect theory: Cumulative representation of uncertainty. Journal of Risk and Uncertainty 5, 297–323 (1992)

Distributed Forward Checking May Lie for Privacy*

Ismel Brito and Pedro Meseguer

IIIA, Institut d'Investigació en Intel.ligència Artificial
CSIC, Consejo Superior de Investigaciones Científicas
Campus UAB, 08193 Bellaterra, Spain
{ismel,pedro}@iiia.csic.es

Abstract. *DisFC* is an *ABT*-like algorithm that, instead of sending the value taken by the high priority agent, it sends the domain of the low priority agent that is compatible with that value. With this strategy, plus the use of sequence numbers, some privacy level is achieved. In particular, each agent knows its value in the solution, but ignores the values of the others. However, the idea of sending the whole compatible domain each time an agent changes its value may cause a privacy loss on shared constraints that was initially overlooked. To solve this issue, we propose $DisFC_{lies}$, an algorithm that works like *DisFC* but it may lie about the compatible domains of other agents. It requires a single extra condition: if an agent sends a lie, it has to tell the truth in finite time afterwards. We prove that the algorithm is sound, complete and terminates. We provide experimental results on the increment in privacy achieved, at the extra cost of more search.

1 Introduction

In the last years, there is an increasing interest for solving constraint satisfaction problems in a distributed form. This has generated a new model, called *DisCSP*, where the information of a *CSP* instance is distributed among several agents but it is never concentrated into a single agent. To solve this new model, new algorithms have appeared that communicate by message passing. Among them, we underline the pioneering *ABT* algorithm [11,12], that has been shown correct and complete.

There are several motivations to solve a *CSP* instance in a distributed form. We can mention the difficulty to collect and move into a single server all the elements of an instance if it is very large, if different formats coexists and the cost of translating them is high. In addition, privacy is a motivation for distributed solving. Many problems appear to be naturally distributed, each part belonging to a different agent. In the solving process, agents desire to keep as private as possible the information they have, and specially they do not want to reveal the values of the solution to other agents.

Although the initial *ABT* was not concerned with privacy issues (agents exchanged their values freely), privacy has been a key aspect for new *DisCSP* solving algorithms. Generally speaking, most distributed algorithms leak some kind of information in the solving process, which can be exploited by some agents to deduce the reserved information of other agents. So far, there are two main approaches to enforce privacy. One considers the use of cryptographic techniques to conceal values and constraints [13,10].

---

* Supported by the Spanish project TIN2005-09312-C03-01.

F. Azevedo et al. (Eds.): CSCLP 2006, LNAI 4651, pp. 93–107, 2007.
© Springer-Verlag Berlin Heidelberg 2007

Alternatively, other authors try to enforce privacy by using different search strategies. Our past work has followed this line, and this paper is a further step on this approach.

In previous work, we proposed *Distributed Forward Checking (DisFC)* [2]. It is an *ABT*-like algorithm that, instead of sending the value of x_i to agent j (assuming i with higher priority than j), it sends the subset of values that j can take which are compatible with the i value. This idea, combined with the formulation of *Partially Known Constraints*, and the use of sequence numbers to conceal the actual value taken by an agent, allow for some degree of privacy. In particular, when a solution is found, each agent knows its own value but ignores the values of other agents. However, we overlooked the effect that sending the whole subset of compatible values may have in constraint privacy. If i has d different values and in the solving process j receives d different compatible subsets, then j knows all the rows of the constraint matrix that i has, but without knowing their position. In the solving process, it is possible to deduce that some positions are discarded for some rows [8]. At the end, agent j may have a non-negligible amount of information about the constraint that i owns, which could be used to break privacy. Nevertheless, computing the set of constraints that are compatible with the information leaked in the solving process requires a significant amount of work (computing all solutions of a *CSP* instance, that is, solving an NP-hard problem).

To prevent this issue, we suggest a new algorithm called $DisFC_{lies}$. It works like standard *DisFC* with a single modification: it may lie in the subsets of compatible values that j may take. Obviously, to keep completeness it has to tell the truth in the values that i truly has. So if i has d values $\{v_1, v_2, \ldots, v_d\}$, $DisFC_{lies}$ works as if i would have $d + k$ values $\{v_1, v_2, \ldots, v_d, v_{d+1}, \ldots, v_{d+k}\}$. We call *true values* as the first d values, while the rest are *false values*. When i takes the true value $v_p, 1 \leq p \leq d$, it sends to agent j the subset of values that are compatible with v_p. When i takes the false value $v_q, d < q \leq d + k$, $DisFC_{lies}$ sends an invented subset of compatible values to j, with the purpose of making more difficult the hypothetical deduction of j on the actual constraint matrix of i. Again, to assure completeness, $DisFC_{lies}$ has to allow all its true values for assignment. As result, this strategy increases the level of privacy at the extra cost of losing performance. This expected result poses a trade-off between efficiency and privacy: enforcing privacy causes to decrease efficiency and vice versa.

In practical terms, what does this mean? First, we have to notice that, even in unsolvable instances (where the major privacy loss occurs), not every agent will have the same level of leaked information. Imagine an instance that contains a single unsolvable subproblem. If the empty nogood is derived exclusively from the interaction of agents in that subproblem, they will have a high level of information about their neighboring constraints, since all possible combinations have been tried. However, agents in other parts of the instance may have less information, if they have reached a consistent assignment with less search. Considering *DisFC* on binary random problems without solution of 16 variables, 10 values per variable and constraint connectivity of 0.4, it may exist one agent that would find 1 matrix compatible with the information leaked, that is, the constraint of the other agent. On the same problems, the new $DisFC_{lies}$ would approximately multiply this number by 2, 20 or 200 when allowing 1, 3 or 5 lies per agent, at the extra cost of incrementing computation and communication costs up to the level of solving problems with 11, 13 or 15 values per variable.

The structure of the paper is as follows. In Section 2 we present the basic concepts used in the paper. In Section 3 we discuss the privacy issues of *DisFC* algorithm. In Section 4 we propose the new algorithm that may lie about the values agents take. In Section 5 we provide experimental results. Finally, in Section 6 we extract some conclusions from this work.

2 Preliminaries

A *Constraint Satisfaction Problem* (*CSP*) involves a finite set of variables, each one taking a value in a finite domain. Variables are related by constraints that impose restrictions on the combinations of values that subsets of variables can take. A *solution* is an assignment of values to variables which satisfies every constraint. Formally, a finite *CSP* is defined by a triple $(\mathcal{X}, \mathcal{D}, \mathcal{C})$, where

- $\mathcal{X} = \{x_1, \ldots, x_n\}$ is a set of n variables;
- $\mathcal{D} = \{D(x_1), \ldots, D(x_n)\}$ is a collection of finite domains; $D(x_i)$ is the initial set of possible values for x_i;
- $\mathcal{C}$ is a set of constraints among variables. A constraint C_i on the ordered set of variables $var(C_i) = (x_{i_1}, \ldots, x_{i_{r(i)}})$ specifies the relation $prm(C_i)$ of the *permitted* combinations of values for the variables in $var(C_i)$. An element of $prm(C_i)$ is a tuple $(v_{i_1}, \ldots, v_{i_{r(i)}})$, $v_i \in D(x_i)$.

A *Distributed Constraint Satisfaction Problem* (*DisCSP*) is a *CSP* where variables, domains and constraints are distributed among automated agents. Formally, a finite *DisCSP* is defined by a 5-tuple $(\mathcal{X}, \mathcal{D}, \mathcal{C}, \mathcal{A}, \phi)$, where $\mathcal{X}$, $\mathcal{D}$ and $\mathcal{C}$ are as before, and

- $\mathcal{A} = \{1, \ldots, p\}$ is a set of p agents,
- $\phi : \mathcal{X} \to \mathcal{A}$ is a function that maps each variable to its agent.

Each variable belongs to one agent. The distribution of variables divides $\mathcal{C}$ in two disjoint subsets, $\mathcal{C}_{intra} = \{C_i | \forall x_j, x_k \in var(C_i), \phi(x_j) = \phi(x_k)\}$, and $\mathcal{C}_{inter} = \{C_i | \exists x_j, x_k \in var(C_i), \phi(x_j) \neq \phi(x_k)\}$, called intraagent and interagent constraint sets, respectively. An intraagent constraint C_i is known by the agent owner of $var(C_i)$, and it is unknown by the other agents. Usually, it is considered that an interagent constraint C_j is known by every agent that owns a variable of $var(C_j)$ [12].

As in the centralized case, a *solution* of a *DisCSP* is an assignment of values to variables satisfying every constraint. *DisCSPs* are solved by the collective and coordinated action of agents $\mathcal{A}$. Agents communicate by exchanging messages. It is assumed that the delay in delivering a message is finite but random. For a given pair of agents, messages are delivered in the order they were sent.

For simplicity purposes, and to emphasize on the distribution aspects, in the rest of the work we assume that each agent owns exactly one variable. We identify the agent number with its variable index ($\forall x_i \in \mathcal{X}, \phi(x_i) = i$). From this assumption, all constraints are inter-agent constraints, so $\mathcal{C} = \mathcal{C}_{inter}$ and $\mathcal{C}_{intra} = \emptyset$. Furthermore, we assume that all constraints are binary. A constraint is written C_{ij} to indicate that it binds variables x_i and x_j.

3   Privacy and DisFC

There are two main concerns about privacy when solving *DisCSP*:

- Privacy of constraints: if agent i is constrained with agent j, i may want to keep private on the part of the constraint known by itself, and the same may occur for j. This generates the *Partially Known Constraints* model (*PKC*) described below.
- Privacy of assignments: agents do not want to reveal the values assigned to their variables to other agents. This is especially relevant for the values of the solution.

3.1   The PKC Model for Constraint Privacy

To enforce constraint privacy, we proposed [2] the *Partially Known Constraints* (*PKC*) model of a *DisCSP* as follows. A constraint C_{ij} is partially known by its related agents. Agent i knows the constraint $C_{i(j)}$ where:

- $var(C_{i(j)}) = \{x_i, x_j\}$;
- $C_{i(j)}$ is specified by three disjoint sets of value tuples for x_i and x_j:
    - $prm(C_{i(j)})$, the set of tuples that i knows to be permitted;
    - $fbd(C_{i(j)})$, the set of tuples that i knows to be forbidden;
    - $unk(C_{i(j)})$, the set of tuples which consistency is not known by i;
- every possible tuple is included in one of the above sets, that is, $prm(C_{i(j)}) \cup fbd(C_{i(j)}) \cup unk(C_{i(j)}) = D_i \times D_j$.

Similarly, agent j knows $C_{(i)j}$, where $var(C_{(i)j}) = \{x_i, x_j\}$. $C_{(i)j}$ is specified by the disjoint sets $prm(C_{(i)j})$, $fbd(C_{(i)j})$ and $unk(C_{(i)j})$ relative to j. Between a constraint C_{ij} and its corresponding partially known constraints $C_{i(j)}$ and $C_{(i)j}$ it holds

$$C_{ij} = C_{i(j)} \otimes C_{(i)j}$$

where $\otimes$ depends on the constraint semantics (see [4] for an example of this). The above definitions satisfy:

- If the combination of values k and l, for x_i and x_j is forbidden in at least one partial constraint, then it is forbidden in the corresponding total constraint.
- If the combination of values k and l, for x_i and x_j is permitted in both partial constraints, then it is also permitted in the corresponding total constraint.

Here, we only consider constraints for which $unk(C_{(i)j}) = unk(C_{i(j)}) = \emptyset$. Then, a partially known constraint $C_{i(j)}$ is completely specified by its permitted tuples and $prm(C_{ij}) = prm(C_{i(j)}) \cap prm(C_{(i)j})$.

For example, let us consider the *n-pieces m-chessboard* problem. Given a set of n chess pieces and a $m \times m$ chessboard, the goal is to put all pieces on the chessboard in such a way that no piece attacks any other. As *DisCSP*, the problem is formulated as,

- Variables: one variable per piece.
- Domains: all variables share the domain $\{1, \ldots, m^2\}$ of chessboard positions (cells are numbered from left to right, from top to bottom).
- Constraints: one constraint between every pair of pieces, from chess rules.
- Agents: one agent per variable.

For instance, we can take $n = 5$ with the multiset of pieces {*queen, castle, bishop, bishop, knight*}, on a 4×4 chessboard, with the variables,

$$x_1 = queen, \ x_2 = castle, \ x_3 = bishop, \ x_4 = bishop, \ x_5 = knight.$$

If agent 1 knows that agent 5 holds a knight, and agent 5 knows that agent 1 holds a queen, this is enough information to develop completely constraint C_{15} by any of them,

$$C_{15} = \{(1, 8), (1, 12), (1, 14), (1, 15), \ldots\}$$

With the *PKC* model, agent 1 does not know which piece agent 5 holds. It only knows how a queen attacks, from which it can develop the constraint,

$$C_{1(5)} = \{(1, 7), (1, 8), (1, 10), (1, 12), \ldots\}$$

Analogously, agent 5 does not know which piece agent 1 holds. Its only information is how a knight attacks, from which it can develop the constraint,

$$C_{(1)5} = \{(1, 2), (1, 3), (1, 4), (1, 5), (1, 6), (1, 8), \ldots\}$$

The whole constraint C_{15} appears as the intersection of these two constraints,

$$C_{15} = C_{1(5)} \cap C_{(1)5} = \{(1, 8), \ldots\}$$

$C_{1(5)}$ does not depend on agent 5. It codifies the way a queen attacks, independently of any other piece.

3.2 Assignment Privacy on *DisFC*

To achieve assignment privacy, we proposed [2] the *Distributed Forward Checking* (*DisFC*) algorithm as follows. In the centralized case, *Forward Checking* (*FC*) [6] filters future domains when the current variable is assigned, removing inconsistent values. *DisFC* extends this idea to the distributed case. It performs an *ABT*-search, with the following differences. When a variable x_i is assigned, instead of sending its value to the connected agent j, it sends to j the part of D_j compatible with its value. Variable x_j will choose a new value consistently with x_i (by selecting its new value from the received filtered domain) but without knowing x_i actual value.

To perform backtracking, variable x_j should know some identifier of the value currently assigned to x_i (otherwise, obsolete backtracking cannot be detected). In *ABT* this identifier is the own value; instead, we propose to use the variable sequence number. Each variable keeps a sequence number that starts from 1 (or some random value), and increases monotonically each time the variable changes its value, acting as a unique identifier for each value. Messages including the sender value replace that value by the sequence number of the sender variable. The agent view of the receiver is composed by the sequence numbers it believes are hold by variables in higher priority agents. Nogoods are formed by variables and their sequence numbers.

DisFC uses both strategies. Each *DisFC* agent sends filtered domains to other agent variables, and it replaces its own value by its sequence number. This allows one agent to exchange enough information with other agents to reach a global consistent solution (or proving that no solution exists) without revealing its own assignment. *DisFC* algorithm performs the same search as *ABT*, with the difference that a constraint is checked by the higher priority agent when sending the filtered domain to the lower priority agent. *DisFC* inherits the correctness and completeness properties of *ABT*. Similar to *ABT*, we can prove that *DisFC* finds a solution or detects inconsistency in finite time.

3.3 DisFC Versions

In the PKC model, if agents i and j are constrained, i knows $C_{i(j)}$ and j knows $C_{(i)j}$, but none knows the total constraint C_{ij}. Assuming this model, there are two versions of *DisFC*. The first proposed was $DisFC_2$ [2]. It consists of a cycle of two phases,

- Phase I. Constraints are directed forming a DAG, and a compatible total order of agents is selected. Then, *DisFC* finds a solution with respect to constraints $C_{i(j)}$, where i has higher priority than j. If no solution is found, the process stops, indicating unsolvable instance.
- Phase II. Constraints and the order of agents are reversed. Now $C_{(i)j}$ are considered, where j has higher priority than i. x_j informs x_i of its filtered domain with respect to x_j value. If the value of x_i is in that filtered domain, i does nothing. Otherwise, i sends a **ngd** message to j, which receives that message and does nothing. Quiescence is detected.

If no **ngd** messages are generated in phase II, the solution provided in phase I also satisfies $C_{(i)j}$, so it is a true solution. Otherwise, phase I restarts. The nogoods generated in phase II are considered by the receiver agents, now with low priority, so they can change their values to find compatible ones. This cycle iterates until a solution is found or the no solution condition is detected. This strategy is correct and complete.

Instead of checking a part of the constraints in phase I and verifying the proposed solution in phase II, Zivan and Meisels proposed that all constraints could be tested simultaneously [14]. Combining this idea with *DisFC*, we obtain the $DisFC_1$ version, that works as follows. An agent has to check all its partially known constraints with both higher and lower priority agents. To do this, an agent has to inform to all its neighbors agents when it takes a new value, and **ngd** messages can go in both directions (from lower to higher as in *ABT* but also from higher to lower). $DisFC_1$ inherits the good properties of *ABT-ASC* [14]. $DisFC_1$ is correct, complete and terminates.

Similar to *DisFC*, $DisFC_1$ agents check consistency with respect to their partial constraints and detect obsolete nogood messages without revealing their assignments. Let $self$ be a generic agent. After an assignment, $self$ informs all constraining agents (with higher and lower priority) via **ok?** messages. Each **ok?** message contains the subset of values for the message recipient that are consistent with $self$'s assignment (filtered domain) and the sequence number corresponding to the $self$'s assignment. In addition, if a conflict exists between $self$'s assignment and a previously received filtered domain from a lower priority agent, a **ngd** message is sent to that agent.

When $self$ receives an **ok?** message from a higher priority agent i, it checks $C_{(i),self}$ looking for a consistent value. $self$ discards those values which are inconsistent with

procedure DisFC-1()
  $myValue \leftarrow$ empty; $end \leftarrow$ false; compute Γ^+, Γ^-;
  CheckAgentView();
  while ($\neg end$) **do**
    $msg \leftarrow$ getMsg();
    switch($msg.type$)
      ok?   : ProcessInfo(msg);
      ngd   : ResolveConflict(msg);
      adl   : SetLink(msg);
      stp, qcc : $end \leftarrow$ true;

procedure CheckAgentView()
  if ($myValue = empty$ **or** $myValue$ eliminated by $myNogoodStore$) **then**
    $myValue \leftarrow$ ChooseValue();
    if ($myValue$) **then**
      $mySeq \leftarrow mySeq + 1$;
      for each $child \in \Gamma^+(self) \cup \Gamma^-(self)$ **do**
          sendMsg:**ok?**($child, mySeq$, compatible($D(child), myValue$));
      for each $child \in \Gamma^+(self)$ such that $\neg (myValue \in MyFilteredDomain[child])$ **do**
          sendMsg:**ngd**($child, self = mySeq \Rightarrow \neg child.Assig$);
    else Backtrack();

procedure ResolveConflict(msg)
  if coherent($msg.Nogood, \Gamma^-(self) \cup \{self\}$) **then**
    CheckAddLink(msg);
    add($msg.Nogood, myNogoodStore$); $myValue \leftarrow$ empty;
    CheckAgentView();
  else if coherent($msg.Nogood, self$) **then**
    SendMsg:**ok?**($msg.Sender, mySeq$, compatible($D(msg.Sender), myValue$));

procedure Backtrack()
  $newNogood \leftarrow$ solve($myNogoodStore$);
  if ($newNogood = $ empty) **then**
    $end \leftarrow$ true; sendMsg:**stp**($system$);
  else
    sendMsg:**ngd**($newNogood$);
    UpdateAgentView(rhs($newNogood$) $\leftarrow$ unknown);
    CheckAgentView();

function ChooseValue()
  for each $v \in D(self)$ not eliminated by $myNogoodStore$ **do**
    if consistent($v, myAgentView[\Gamma^-]$) **then return** (v);
    else add($x_j = val_j \Rightarrow self \neq v, myNogoodStore$); /*$v$ is inconsistent with x_j's value */
  return (empty);

procedure UpdateAgentView($newAssig$)
  add($newAssig, myAgentView$);
  for each $ng \in myNogoodStore$ **do**
    if $\neg$Coherent(lhs(ng), $myAgentView$) **then** remove($ng, myNogoodStore$);

procedure SetLink(msg)
  add($msg.sender, \Gamma^+(self)$);
  sendMsg:**ok?**($msg.sender, myValue$);

procedure CheckAddLink(msg)
  for each ($var \in$ lhs($msg.Nogood$))
    if ($var \notin \Gamma^-(self)$) **then**
      sendMsg:**adl**($var, self$);
      add($var, \Gamma^-(self)$); UpdateAgentView($var \leftarrow varValue$);

Fig. 1. The $DisFC_1$ algorithm for asynchronous backtracking search

higher priority agents. If no consistent value is found, $self$ backtracks solving conflicts in $myNogoodStore$, as in ABT, sending a **ngd** message. When $self$ receives an **ok?** message from a lower priority agent j, it checks $C_{self(j)}$. If the assignments of $self$ and j are not consistent, $self$ sends a **ngd** message informing to j that its assignment is not valid for $self$'s assignment. Otherwise, $self$ does nothing. **ngd** messages are processed in the same way, no matter if they come from higher or lower priority agents.

The code of $DisFC_1$ appears in Figure 1. This code is concurrently executed by each agent. In $myAgentView$, each agent stores the sequence number received from its neighboring agents. In $myNogoodStore$, each agent stores received nogoods from higher and lower agents. In $myFilteredDomains$, each agent saves the last filtered domains received from its (higher and lower) neighbors. Γ^- refers to agents related to $self$ with higher priority, while Γ^+ refers to agents related to $self$ with lower priority.

Agents exchange five kind of messages: **ok?**, **ngd**, **adl**, **stp** and **qcc**. The meaning of **ok?** and **ngd** messages has been described above. **adl** has the same uses as in ABT and $DisFC_2$: to connect unrelated agents. An extra agent called $system$ controls the termination of the algorithm by using **stp** and **qcc** messages. When an agent finds inconsistency it sends an **stp** message to $system$. When $system$ receives an **stp** message from one agent or detects quiescence in the network (i.e. no message has traveled through the network in the last t_{quies} units of time), $system$ sends messages to all agents informing them to finish the search. In former case, $system$ sends **stp** messages to all agents, which is to say that the problem is unsolvable. In latter case, $system$ sends **qcc** messages, which is to say that the problem has at least one solution which is given by the current variables' assignments. Quiescence state can be detected by specialized algorithms [5].

3.4 Breaking Privacy

Comparing $DisFC$ with ABT, the basic difference is as follows. If agent i is constrained with j and i has higher priority, instead of sending the actual value of i to j, it sends the subset of D_j that is compatible with the actual value of i. After reception, j does not know the actual value of i, but it knows a complete row of $C_{i(j)}$ without knowing its position in the matrix. As search progresses, j may store new rows of $C_{i(j)}$. At the end, j has a subset of rows without knowing their position. In addition, some search episodes (changing from phase I to phase II in $DisFC_2$, nogood messages from high to low priority agents in $DisFC_1$) may reduce the number of acceptable positions for a particular row [8]. With all this, j may construct a CSP instance where the variables are the rows, their domains are the acceptable positions, under the constraints that two different rows cannot go to the same position and every row must get a position. Computing all solutions of this instance we obtain all matrixes which are compatible with the information obtained from the search. Of them, one is $C_{i(j)}$. So to break privacy, all solutions of this CSP instance have to be computed (an NP-hard task). In practice, solving this instance requires significant effort and in some cases subsumption testing is required.

4 *DisFC* May Lie

To enhance privacy in *DisFC* we propose that agents could lie. Instead of sending true rows of $C_{i(j)}$, the algorithm may send true and *false* rows. Each false row represents a lie. False rows will make much more difficult the hypothetical reconstruction of $C_{i(j)}$ by agent j, but it has to be done keeping the soundness and completeness of the algorithm. This idea can be formalized as follows. If i has d values $D_i = \{v_1, v_2, \ldots, v_d\}$, it is assumed that i has an extended domain $D'_i = \{v_1, v_2, \ldots, v_d, v_{d+1}, \ldots, v_{d+k}\}$ of $d + k$ values. We call *true values* the first d values, while the rest are *false values*. When i assigns the true value $v_p, 1 \leq p \leq d$, it sends to agent j the subset of values that are compatible with v_p (that is, a true row of $C_{i(j)}$). When i assigns the false value $v_q, d < q \leq d + k$, it sends an invented subset of compatible values to j (that is, a row which does not exist in $C_{i(j)}$). The only concern that an agent must have after assigning a false value is that it must tell the truth (assign a true value or perform backtracking if no more true values are available) in finite time. The point is that no solution could be based on a false value, so assignments including false values have to be removed in finite time (in fact, in a shorter time than required to detect quiescence).

4.1 The *DisFC_{lies}* Algorithm

DisFC_1 offers a better platform for privacy than *DisFC_2*, because it has no synchronization points between phases. For this reason, we implement the lies idea on top of *DisFC_1* (although it can also be implemented on top of *DisFC_2*).

We call *DisFC_{lies}* the new version of *DisFC_1* where agents may exchange false pruned domains. *DisFC_{lies}* appears in Figure 2. It includes most of the procedures, functions and data structures of *DisFC_1*, and uses the same types of messages. Each agent has a local clock to control when it has to tell the truth after a lie. In the structure $FalseDomains$, each agent puts away the false domains that it will send to its neighbors for each false value the agent's variable can take. $D_{true}(self)$ is the set of true values for $self$, while $D_{false}(self)$ is the set of its false values. $D(self)$ is the union of these two sets.

In the main procedure, $self$ first initializes its data structures and generates the false domain that it will sent for each false value. Secondly, $self$ assigns a value to its variable by invoking the function $CheckAgentView$. This value may be false or not. Then, $self$ enters in a loop, where incoming messages are received and processed. This loop ends, and therefore the algorithm, when $self$ receives either an **stop** or a **qcc** message from $system$. This is a special agent that handles these messages in the same way it did in *DisFC_1*. If $self$ ends the search because a **qcc** message, it means that a problem has at least one solution, otherwise, the problem is unsolvable. Quiescence state can be detected by specialized algorithms [5]. However, in order to assure the completeness and soundness of the algorithm, the time t_{quies} required by $system$ to assure quiescence in the network (i.e no message has traveled through the network within the last t_{quies} units of time) must be larger than t_{lies}, the maximum time agents may wait until rectifying their lies, thus $t_{lies} < t_{quies}$.

In the following, we prove that *DisFC_{lies}* is sound, complete and terminates.

procedure DisFC$_{lies}$ ( )
  $myValue \leftarrow$ empty; $end \leftarrow$ false; compute Γ^+, Γ^-; $tsaytrue \leftarrow 0$;
  for each $value \in D_{false}(self)$ **do**
    for each $neig \in \Gamma^+(self) \cup \Gamma^-(self)$ **do** generate $FalseDomain[value][neig]$;
  CheckAgentView();
  while ($\neg end$) **do**
    $msg \leftarrow$ getMsg();
    switch($msg.type$)
      ok? : ProcessInfo(msg);
      ngd : ResolveConflict(msg);
      adl : SetLink(msg);
      stp, qcc : $end \leftarrow$ true;
      if ($value \in D_{false}(self)$) **and** (gettime() $\geq tsaytrue$) **then** TakeATrueValue();
procedure CheckAgentView()
  if ($myValue =$ empty **or** $myValue$ eliminated by $myNogoodStore$) **then**
    $myValue \leftarrow$ ChooseValue ( );
  if ($myValue$) **then**
    $mySeq \leftarrow mySeq + 1$;
    if ($myValue \in D_{false}(self)$) **then**
      for each $neig \in \Gamma^+(self) \cup \Gamma^-(self)$ **do**
        sendMsg:**ok?**($neig, mySeq, FalseDomain[myValue][neig]$);
      $tsaytrue \leftarrow$ gettime() $+ t_{lies}$;   /* $t_{lies} < t_{quies}$ */
    else
      for each $neig \in \Gamma^+(self) \cup \Gamma^-(self)$ **do**
        sendMsg:**ok?**($neig, mySeq,$ compatible($D(neig), myValue$));
      for each $child \in \Gamma^+(self)$ such that $\neg (myValue \in MyFilteredDomain[child])$ **do**
        sendMsg:**ngd**($child, self = mySeq \Rightarrow \neg child.Assig$);
      $tsaytrue \leftarrow 0$;
  else Backtrack();
procedure ResolveConflict(msg)
  if coherent($msg.Nogood, \Gamma^-(self) \cup \{self\}$) **then**
    CheckAddLink(msg);
    add($msg.Nogood, myNogoodStore$); $myValue \leftarrow$ empty;
    CheckAgentView();
  else if coherent($msg.Nogood, self$) **then**
    if ($myValue \in D_{false}(self)$) **then**
      sendMsg:**ok?**($neig, mySeq, FalseDomain[myValue][neig]$);
    else
      SendMsg:**ok?**($msg.Sender, mySeq,$ compatible($D(msg.Sender), myValue$));
procedure TakeATrueValue()
  $tsaytrue \leftarrow 0$; $myValue \leftarrow$ ChooseATrueValue ( );
  if ($myValue$) **then**
    $mySeq \leftarrow mySeq + 1$;
    for each $neig \in \Gamma^+(self) \cup \Gamma^-(self)$ **do**
      sendMsg:**ok?**($neig, mySeq,$ compatible($D(neig), myValue$));
    for each $child \in \Gamma^+(self)$ such that $\neg (myValue \in MyFilteredDomain[child])$ **do**
      sendMsg:**ngd**($child, self = mySeq \Rightarrow \neg child.Assig$);
  else Backtrack();
function ChooseATrueValue ( )
  for each $v \in D_{true}(self)$ not eliminated by $myNogoodStore$ **do**
    if consistent($v, myAgentView[\Gamma^-]$) **then return** (v);
    else add($x_j = val_j \Rightarrow self \neq v, myNogoodStore$); /*$v$ is inconsistent with x_j's value */
  return (empty);

Fig. 2. The $DisFC_{lies}$ algorithm for asynchronous backtracking search. Missing procedures/functions appear in Figure 1.

4.2 Theoretical Results

Lemma 1. *When $DisFC_{lies}$ finds a solution, the last filtered domain received by agent i from agent j corresponds to a (true) row in the partial constraint matrix $C_{i(j)}$.*

Proof. For $DisFC_{lies}$ the current variables' assignments are a solution if no constraint is violated and network has reached quiescence. Let us assume that $DisFC_{lies}$ reports a solution in which variable x_i takes a false value. So the last filtered domains sent by agent i are false too. However, $DisFC_{lies}$ requires that, after lying, an agent must rectify in finite time. That is, assigning a true value and sending to its neighbors the true filtered domains, or performing backtrack. So, at least one **ok?** message or a **ngd** message has traveled through the network after i lied, in contradiction with the initial assumption that the network had reached quiescence. Therefore, the solution condition cannot be reached unless true filtered domains are sent in the last messages from any agent.   □

Proposition 1. *$DisFC_{lies}$ is sound.*

Proof. If a solution is claimed, we have to prove that current agents' assignments satisfy their partial constraints. Lemma 1 shows that if $DisFC_{lies}$ reports a solution the last variables's assignments correspond to true values. Therefore, one can prove that $DisFC_{lies}$ is sound by using the same arguments to prove that $DisFC_1$ is sound.

Let us assume quiescence in the network. If the current assignment is not a solution, there exists at least one partial constraint that is violated by agent j. In that case, agent j has sent a **ngd** message to agent i, the closest agent involved in the conflict. This **ngd** is either discarded as obsolete or accepted as valid by agent i. If the message is discarded, it means that some message has not yet reached its recipient, which breaks our assumption of quiescence in the network. If the message is valid, i has to find a new consistent values, which will produce several **ok?** messages or one new **ngd** message, which again breaks our assumption of quiescence in the network.   □

Proposition 2. *$DisFC_{lies}$ is complete.*

Proof. Considering only nogoods based on true values, we can prove that $DisFC_{lies}$ is complete by using the same arguments to prove that $DisFC_1$ is complete. Since nogoods resulting from an **ok?** message are redundant with respect to the partial constraint matrixes, and the additional nogoods are generated by logical inference, the empty nogood cannot be inferred if the problem is solvable.

Let us prove that $DisFC_{lies}$ cannot infer inconsistency based on false values if the problem is solvable. Suppose that agent j detects inconsistency because a lie introduced by agent i. We know that j detects inconsistency when it infers an empty nogood. Besides, we know that the left-hand side of the nogoods (justifications of forbidden values) stored by j is either empty or includes agents with higher priority than j. Since we assume that inconsistency discovered by j is based on the false value of i, i is before j in the agents' ordering and there is at least one nogood stored by j including i in its left-hand side. Therefore, when j finds no consistent value, it has to send a backtracking messages to i, which breaks our assumption that j derives an empty nogood.   □

Lemma 2. *$DisFC_{lies}$ agents will not store indefinitely nogoods based on false values.*

Proof. Let us assume that a false nogood (i.e. a nogood including an agent with a false value) will be stored indefinitely by an agent. In that case, the lying agent cannot change its variable's assignment, otherwise the nogood will become obsolete and, therefore, deleted by the holder agent. But a lying agent *must* tell the truth in finite time. So, in finite time, the agent storing the false nogood will be informed of a new true value, the false nogood will become obsolete and, therefore, it will be deleted by the holder agent. This breaks our assumption that the false nogood lasts forever.   □

Proposition 3. *DisFC$_{lies}$ terminates.*

Proof. By Lemma 2, nogoods based on false values are discarded in finite time. About nogoods based on true values, $DisFC_{lies}$ performs the same treatment as $DisFC_1$. Since $DisFC_1$ terminates in finite time, $DisFC_{lies}$ also terminates in finite time.   □

Proposition 4. *If a DisFC$_{lies}$ agent detects inconsistency, every agent directly connected with it has received d true rows.*

Proof. Let i be that agent. If i finds the empty nogood, it means that there is a nogood for every true value of i. These nogoods have an empty left-hand side (otherwise, i could not deduce the empty nogood). So they have been produced as result of **ngd** messages coming from the lower priority agents. Therefore, every possible true value of i has been taken, so i has sent to its neighbors d true rows.   □

4.3   Privacy Improvements of *DisFC$_{lies}$*

The inclusion of false values has two direct consequences. First, agent j may receive false rows of $C_{i(j)}$. Then j has more difficulties to reconstruct $C_{i(j)}$, since it is uncertain whether some received rows truly belong to $C_{i(j)}$ or not. Second, this strategy decreases performance, because any computation that includes a false assignment will not produce any solution, so it is a wasted effort, only useful for privacy purposes.

For a solvable instance, Lemma 1 shows that the last assignments correspond to true values. So, agent j knows that the last message from i correspond to a true assignment, and it contains a true row of $C_{i(j)}$. Agent j cannot discriminate whether previous assignments are true or false, so it cannot include the rows of these messages when trying to compute $C_{i(j)}$. So j knows a single row of $C_{i(j)}$ but it does not know its location. The number of different constraint matrixes compatible with this information is approximately $d \cdot 2^{(d^2-d)}$ (d, the number of possible locations for the true row, times $2^{(d^2-d)}$, the number of compatible matrixes when d elements are known). This is a big difference with the approach without lies, where all received rows truly belong to $C_{i(j)}$.

For an unsolvable instance, Proposition 4 shows that every agent j directly connected with the agent i that detects inconsistency would have received d true rows. In addition, since all possibilities have been tried, they have received $d + k$ rows (observe that j cannot receive more than $d + k$ rows). Assuming that j has received $d + k$ *different* rows, if j wants to compute $C_{i(j)}$, it has to select d rows, take them as true rows and solve the corresponding *CSP*. j has to repeat this process $\binom{d+k}{d}$ times, that is, once for each different subset of d rows. This increases the number of *CSP*s to solve, in order to compute the matrixes compatible with the leaked information. However, j may have received less than $d + k$ *different* rows. In that case, j considers that some rows are repeated. If there is no way to identify repeated rows, in addition to the previously described combinations, we have to consider each possible row as possible repeated, increasing greatly the number of *CSP* instances to solve. As consequence, the privacy level of the solving process is improved.

5   Experimental Results

In this Section, we compare the performance of $DisFC_1$ and $DisFC_{lies}$ solving instances of binary random classes. A binary random class is defined by $\langle n, d, p_1, p_2 \rangle$, where n is the number of variables, d the number of values per variable, p_1 the network *connectivity* (the ratio of existing constraints) and p_2 the constraint *tightness* (the ratio of forbidden value pairs). We solved instances of the class $\langle 15, 10, 0.4, p_2 \rangle$ with varying tightness (p_2) between 0.1 to 0.9 in increments of 0.1. For creating these instances in PKC, first we generate random instances and then we split the forbidden tuples of each constraint between its two partial constraints.

We consider three versions of $DisFC_{lies}$ that differ from each other in the number of false values that their agents add to initial domains: 1, 3 and 5 false values. Results for all algorithms were produced using a simulator, in which agents are individual processes. Agents are activated randomly. When an agent takes a value, it chooses between true and false values with probability 0.5. t_{lies} is randomly chosen between 1 and 99 internal units of time. Messages are processed by packets, as described in [3].

Algorithmic performance is evaluated by communication effort, computation cost and privacy of constraints. Communication effort is measured by the total number of exchanged messages (msg). Computation cost is measured by the number of non-concurrent constraint checks ($nccc$) [7]. Privacy of constraints is measured by the number of constraint matrixes consistent with the information exchanged among agents. Generally, lower priority agents work more than higher priority ones, therefore they reveals more information than higher priority ones. Thus, we report the minimum (min), median (med) and average (avg) of the numbers of constraint matrixes that are consistent with information exchanged among agents.

Figure 3 shows the computation and communication costs. In both plots, results are averaged on 100 instances. In terms of computation cost, we observe that $DisFC_{lies}$ is more costly than $DisFC$, and the cost increases with the number of allowable lies. The differences between algorithms are greater at the difficulty peak ($p_2 = 0.6$). Except for $p_2 = 0.5$, $DisFC_{lies}(5)$ always requires more $nccc$ than the others, while $DisFC$ performs the lowest number of $nccc$. Similar results appear for communication costs.

Table 1 contains the values of parameters min, med and avg to measure the privacy of constraints. Larger values mean higher privacy. The critical privacy occurs when the number of constraint matrixes is 1 (at least one agent knows exactly the partial constraint matrix of one of its constraining agents). Regarding privacy of constraints in $DisFC_1$, the values of min and med decrease when p_2 increases. Actually, in problems with constraint tightness greater than 0.4, at least one agent can infer exactly the partial constraint of one of its constraining agents (see column min). From med values in

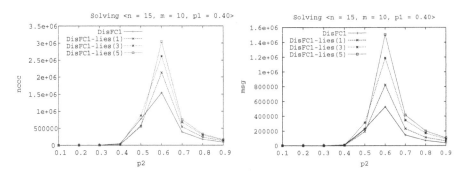

Fig. 3. Computation and communication cost of $DisFC$ and versions of $DisFC_{lies}$

unsolvable instances ($p_2 \geq 0.6$), we conclude that approximately in half of partial constraint matrixes all rows are revealed during search since 10^6 is close to $10! = 3.6 \times 10^6$ (the number of permutations of 10 rows). In terms of avg, higher privacy loss occurs at the complexity peak ($p_2 = 0.6$).

Regarding privacy of constraints in $DisFC_{lies}$, we notice the following. In solvable instances ($0.1 \leq p_2 \leq 0.5$), $DisFC_{lies}$ versions achieve the same level of privacy for min, med and avg, no matter the number of allowable lies. This occurs since each agent can only assure that the last filtered domain received from another agent truly corresponds to a row in the partial constraint matrix of that agent (see Lemma 1), which is independent to the number of false values that agents may have. In terms of min and med, $DisFC_{lies}$ versions are more private than $DisFC_1$. In unsolvable instances, $DisFC_{lies}$ versions have different level of privacy when considering min. $DisFC_{lies}(5)$ is one and two orders of magnitude more private than $DisFC_{lies}(3)$ and $DisFC_{lies}(1)$, respectively. $DisFC_{lies}(1)$ is the least private of these three algorithms although it is more private than $DisFC$. $DisFC_{lies}$ versions are equally private with respect to med and avg. For these parameters, $DisFC_{lies}$ versions are more private than $DisFC_1$.

Table 1. Privacy of constraints measured by the minimum (min), median (med) and average (avg) of the numbers of consistent constraint matrixes. Averaged on 10 instances.

	$DisFC_1$			$DisFC_{lies}(1)$			$DisFC_{lies}(3)$			$DisFC_{lies}(5)$		
p_2	min	med	avg	min	med	avg	min	med	avg	min	med	avg
0.1	10^{23}	10^{28}	10^{29}	10^{27}	10^{28}	10^{28}	10^{27}	10^{28}	10^{28}	10^{27}	10^{28}	10^{28}
0.2	10^{23}	10^{27}	10^{29}	10^{27}	10^{28}	10^{28}	10^{27}	10^{28}	10^{28}	10^{27}	10^{28}	10^{28}
0.3	10^{16}	10^{24}	10^{29}	10^{27}	10^{28}	10^{28}	10^{27}	10^{28}	10^{28}	10^{27}	10^{28}	10^{28}
0.4	10^7	10^{14}	10^{28}	10^{27}	10^{28}	10^{28}	10^{27}	10^{28}	10^{28}	10^{27}	10^{28}	10^{28}
0.5	1	10^9	10^{25}	10^{27}	10^{28}	10^{28}	10^{27}	10^{28}	10^{28}	10^{27}	10^{28}	10^{28}
0.6	1	10^6	10^9	**3.3**	10^{30}	10^{29}	**20**	10^{30}	10^{29}	**221**	10^{30}	10^{29}
0.7	1	10^6	10^{12}	**2**	10^{30}	10^{29}	**10.7**	10^{30}	10^{29}	**163**	10^{30}	10^{29}
0.8	1	10^6	10^{10}	**2.3**	10^{30}	10^{29}	**50.3**	10^{30}	10^{29}	**270**	10^{30}	10^{29}
0.9	1	10^6	10^{10}	**3.3**	10^{30}	10^{29}	**25.3**	10^{30}	10^{29}	**426**	10^{30}	10^{29}

6   Conclusion

From this work we can extract the following conclusions. First, lying is a suitable strategy to enhance privacy in $DisCSP$ solving. We have presented $DisFC_{lies}$, a new version of the $DisFC$ algorithm that may tell lies, sending false compatible domains to neighbor agents. The unique extra condition is that, after a lie, the lying agent has to tell the truth in finite time, lower than t_{quies}. We have proved that this algorithm is correct, complete and terminates. Second, we have shown analytical and experimentally that this idea effectively enhances constraint privacy in the PKC model, because it increases the number of partially known constraint matrixes that are compatible with the leaked information of the solving process. And third, although solving $DisCSP$ lying is more costly than solving it without lies, experiments show that the extra cost required is not unreachable. It is clear that any strategy used to conceal information will have an extra cost, and this approach is not an exception. We believe that this approach could be useful for those applications with high privacy requirements.

References

1. Bessiere, C., Brito, I., Maestre, A., Meseguer, P.: The Asynchronous Backtracking without adding links: a new member in the ABT family. Artifical Intelligence 161, 1–2, 7–24 (2005)
2. Brito, I., Meseguer, P.: Distributed Forward Checking. In: Rossi, F. (ed.) CP 2003. LNCS, vol. 2833, pp. 801–806. Springer, Heidelberg (2003)
3. Brito, I., Meseguer, P.: Synchronous, Asynchronous and Hybrid algorithms for DisCSP. In: CP-2004, Workshop on Distributed Constraint Reasoning (2004)
4. Brito, I., Meseguer, P.: Distributed Stable Matching Problems with Ties and Incomplete Lists. In: Benhamou, F. (ed.) CP 2006. LNCS, vol. 4204, pp. 675–680. Springer, Heidelberg (2006)
5. Chandy, K., Lamport, L.: Distributed Snapshots: Determining Global States of Distributed Systems. ACM Trans. Computer Systems 3(2), 63–75 (1985)
6. Haralick, R., Elliot, G.: Increasing Tree Search Efficiency for Constraint Satisfaction Problems. Artificial Intelligence 14, 263–313 (1980)
7. Meisels, A., Kaplansky, E., Razgon, I., Zivan, R.: Comparing Performance of Distributed Constraint Processing Algorithms. In: AAMAS-02 Workshop on Distributed Constraint Reasoning, pp. 86–93 (2002)
8. Meisels, A., Zivan, R.: Personal communication (2006)
9. Silaghi, M.C., Sam-Haroud, D., Faltings, B.: Asynchronous Search with Aggregations. In: Proc. of the AAAI-2000, pp. 917–922 (2000)
10. Silaghi, M.C.: Solving a distributed CSP with cryptographic multi-party computations, without revealing constraints and without involving trusted servers. In: IJCAI 2003, Workshop on Distributed Constraint Reasoning (2003)
11. Yokoo, M., Durfee, E., Ishida, T., Kuwabara, K.: Distributed Constraint Satisfaction for Formalizing Distributed Problem Solving. In: Proc. of the 12th. DCS, pp. 614–621 (1992)
12. Yokoo, M., Durfee, E., Ishida, T., Kuwabara, K.: The Distributed Constraint Satisfaction Problem: Formalization and Algorithms. IEEE Trans. Knowledge and Data Engineering 10, 673–685 (1998)
13. Yokoo, M., Suzuki, K., Hirayama, K.: Secure Distributed Constraint Satisfaction: Reaching Agreement without Revealing Private Information. In: Van Hentenryck, P. (ed.) CP 2002. LNCS, vol. 2470, pp. 387–401. Springer, Heidelberg (2002)
14. Zivan, R., Meisels, A.: Asynchronous Backtracking for Asymmetric DisCSPs. In: IJCAI 2005 Workshop on Distributed Constraint Reasoning (2005)

Solving First-Order Constraints in the Theory of the Evaluated Trees

Thi-Bich-Hanh Dao[1] and Khalil Djelloul[2]

[1] Laboratoire d'Informatique Fondamentale d'Orléans, France
[2] Faculty of Computer Science, University of Ulm, Germany

Abstract. We present in this paper a first-order extension of the solver of Prolog III, by giving not only a decision procedure, but a full first-order constraint solver in the theory T of the evaluated trees, which is a combination of the theory of finite or infinite trees and the theory of the rational numbers with addition, subtraction and a linear dense order relation. The solver is given in the form of 28 rewriting rules which transform any first-order formula φ into an equivalent disjunction ϕ of simple formulas in which the solutions of the free variables are expressed in a clear and explicit way. The correctness of our algorithm implies the completeness of a first-order theory built on the model of Prolog III.

1 Introduction

The algebra of finite or infinite trees plays a fundamental role in computer science: it is a model for data structures, program schemes and program executions. As early as 1976, G. Huet proposed an algorithm for unifying infinite terms, that is solving equations in that algebra [13]. B. Courcelle has studied the properties of infinite trees in the scope of recursive program schemes [7]. A. Colmerauer has described the execution of Prolog II, III and IV programs in terms of solving equations and disequations in that algebra [4,3,1].

The unification of finite terms, i.e. solving conjunctions of equations in the theory of finite trees has first been studied by A. Robinson [24]. Some algorithms with better complexities have been proposed after by M.S. Paterson and M.N.Wegman [22] and A. Martelli and U. Montanari [21]. A good synthesis on this field can be found in the paper of J.P. Jouannaud and C. Kirchner [15]. Solving conjunctions of equations on infinite trees has been studied by G. Huet [13], by A. Colmerauer [5] and by J. Jaffar [14]. Solving conjunctions of equations and disequations on finite or infinite trees has been studied by H.J. Burckert [2] and A. Colmerauer [4]. An incremental algorithm for solving conjunctions of equations and disequations on rational trees has then been proposed by V.Ramachandran and P. Van Hentenryck [23].

On the other hand, M.J. Maher has axiomatized the theory of finite or infinite trees and showed its completeness using a decision procedure which transforms any first-order formula into a Boolean combination of quantified conjunctions of atomic formulas [19]. A much more general decision procedure was recently given by K. Djelloul in the frame of decomposable theories [12].

F. Azevedo et al. (Eds.): CSCLP 2006, LNAI 4651, pp. 108–123, 2007.

We have then extended Maher's theory of finite or infinite trees by giving a complete first-order axiomatization of the evaluated trees which are a combination of finite or infinite trees with construction operations and the rational numbers with addition, subtraction and a linear dense order relation [9]. This theory, denoted by T, reflects essentially to Prolog III which has been modeled by A. Colmerauer [3] using a combination of trees and rational numbers. Nevertheless, the solver of Prolog III is not able to solve arbitrary quantified first-order constraints built on a combination of trees and rational numbers.

A first attempt of an extension of the solver of Prolog III was given in [11]. It consists in a decision procedure which for every proposition (formula without free variables) gives either true of false in T. Unfortunately, this decision procedure is not able to solve first-order constraints having free variables. In fact, it does not warrant that the solutions of the free variables of a solved formula are expressed in a clear and explicit way and can even produce, starting from a formula φ which contains free variables, an equivalent solved formula ϕ having free variables but being always false (or always true) in T. The appropriate solved formula of φ in this case should be the formula *false* (or the formula *true*) instead of ϕ.

Much more elaborated algorithms are then needed, specially when we want to induce solved formulas expressing solutions of complex first-order constraint satisfaction problems in T. Of course, our goal in these kinds of problems is not only to know if there exist solutions or not, but to express these solutions in the form of a solved formula which is either the formula *true* (i.e. the problem is always satisfiable) or the formula *false* (i.e. the problem is always unsatisfiable) or a simple first-order formula which is neither equivalent to *true* nor to *false* and where the solutions of the free variables are expressed in a clear and explicit way. Algorithms which are able to produce such a formula are called *first-order constraint solvers*.

We present in this paper, not only a decision procedure, but a full first-order constraint solver which gives clear and explicit solutions for any first-order constraint satisfaction problem in T. Our solver is not simply a combination of an algorithm over trees with one over rational numbers, but a powerful mechanism to solve mixed constraints. It includes full systems of typing deduction and constraint simplification and propagation. One of the major difficulties in this work resides in the fact that (i) the theory of finite or infinite trees does not accept full elimination of quantifiers, (ii) every algorithm deciding propositions in the theory of finite or infinite trees has a non-elementary complexity [25] and (iii) the function symbols $+$ and $-$ of T have two different behaviors whether they are applied on trees or rational numbers. For example $+(1, 1)$ is the rational number 2, while $+(1, f_0)$ is the tree whose root is labeled $+$ and whose sons are 1 and the tree's constant f_0.

One of the practical applications of our solver is a powerful extension of the internal solver of Prolog III by allowing the user to handle general first-order constraints and solve them in T. The solver will then give the solutions of the free variables in all the models of T and present them in a clear and explicit way. As far as we know, this is the first algorithm which is able to do a such work. Solving quantified constraints over trees and rational numbers can also be used in PSPACE-complete decision problems from areas such as planning under uncertainty, adversary game playing, and model checking. For example, in game playing we may want to find a winning strategy *for all* possible moves of the opponent. In a manufacturing problem it may be required that a configuration must be possible *for all* possible sequences of user choices. Finally, when planning in a safety critical environment, such as a nuclear power station, we require that an action is possible *for every* eventuality.

The paper is organized in four sections followed by a conclusion. This introduction is the first section. In Section 2 we present the theory of the evaluated trees and introduce an example of a complex constraint in this theory. In Section 3, we define the notions of basic formulas, blocks and solved blocks in T which are particular conjunctions of atomic formulas. We end this section by showing that every quantified solved block can be decomposed in three embedded sequences of quantifications having particular properties which enable us to eliminate some quantifiers. In Section 4, we present the working formulas, the general solved formulas and the algorithm of constraint solving in T. The algorithm is presented in the form of 28 rewriting rules and transforms an initial working formula of depth d into a final working formula of depth less than or equal to three. The main idea behind this algorithm consists in (1) a top-down simplification and propagation of constraints. In each level, quantified blocks are locally solved, decomposed and then propagated to the embedded sub-formulas. Inconsistent sub-formulas are also removed (2) a bottom-up elimination of quantifiers and working formulas' depth decrease using distribution. The disjunction ϕ of general solved formulas extracted from the final working formula is either the formula *false* or *true* or a formula having at least one free variable, being equivalent neither to *false* nor to *true* in T, and where the solutions of the free variables are expressed in a clear and explicit way. We end this section by giving an example of a constraint having two free variables but being always false in T.

The algorithm represented by a set of rewriting rules and the general solved formulas are our main contribution in this paper. The expressiveness and clearness of the solutions of the free variables in the final solved formula are our main goal in this work.

2   Theory T of Evaluated Trees

2.1   Preliminaries

Let F be an infinite set of *function symbols* containing the symbols $+$, $-$, 0 and 1. To each element of F is associated a non-negative integer, its *arity*. The arities of $+$, $-$, 0 and 1 are respectively 2, 1, 0 and 0. Let $R = \{<, num, tree\}$ be the set of relation symbols, of respective arities 2, 1 and 1. Let V be an infinite countable set of *variables*. A *term* is an expression of the form x or $ft_1 \ldots t_n$ where $n \geq 0$, f an n-ary symbol in F and the t_i's are shorter terms. A *formula* is an expression of the forms:

$s = t$, $rt_1..t_n$, *true*, *false*, $\neg(\varphi)$, $(\varphi \wedge \psi)$, $(\varphi \vee \psi)$, $(\varphi \rightarrow \psi)$, $(\varphi \leftrightarrow \psi)$, $\exists x \varphi$, $\forall x \varphi$, where $x \in V$, s, t and the t_i's are terms, r is an n-ary relation symbol in R and φ and ψ are shorter formulas. The first four forms are called atomic. An occurrence of a variable x in a formula is *bound* if it occurs in a sub-formula of the form $(\exists x \varphi)$ or $(\forall x \varphi)$. It is *free* otherwise. The *free variables* of a formula are those which have at least a free occurrence in the formula. For each formula φ, we denote by $\text{var}(\varphi)$ the set of all the free variables of φ. Let $\bar{x} = x_1 \ldots x_n$ and $\bar{y} = y_1 \ldots y_n$ be two vectors of variables of the same length. The empty vector is denoted by ε. Let φ and $\varphi(\bar{x})$ be formulas. We write

$$\exists \bar{x} \, \varphi \quad \text{for } \exists x_1 ... \exists x_n \, \varphi,$$
$$\forall \bar{x} \, \varphi \quad \text{for } \forall x_1 ... \forall x_n \, \varphi,$$
$$\exists ? \bar{x} \, \varphi(\bar{x}) \text{ for } \forall \bar{x} \forall \bar{y} \, \varphi(\bar{x}) \wedge \varphi(\bar{y}) \rightarrow \bigwedge_{i \in \{1,...,n\}} x_i = y_i,$$
$$\exists ! \bar{x} \, \varphi \quad \text{for } (\exists \bar{x} \, \varphi) \wedge (\exists ? \bar{x} \, \varphi).$$

Semantically, the new quantifiers $\exists ?$ and $\exists !$ simply mean "at most one" and "one and only one".

2.2   Axiomatization of T

Let a be a positive integer and let $t_1, ..., t_n$ be terms. Let us denote by:

- $t_1 < t_2$, the term $< t_1 t_2$,
- $t_1 + t_2$, the term $+ t_1 t_2$,
- $t_1 + t_2 + t_3$, the term $+ t_1 (+ t_2 t_3)$,
- $0 t_1$, the term 0,
- $a t_1$, the term $\underbrace{t_1 + \cdots + t_1}_{a}$,
- $-a t_1$, the term $\underbrace{(-t_1) + \cdots + (-t_1)}_{a}$.

The theory T of the evaluated trees is the set of first-order propositions of the following forms:

1   $\forall \bar{x} \forall \bar{y} \, ((tree \, f\bar{x}) \wedge (tree \, f\bar{y}) \wedge f\bar{x} = f\bar{y}) \rightarrow \bigwedge_i x_i = y_i$,
2   $\forall \bar{x} \forall \bar{y} \, f\bar{x} = g\bar{y} \rightarrow num \, f\bar{x} \wedge num \, g\bar{y}$,
3   $\forall \bar{x} \forall \bar{y} \, ((\bigwedge_{i \in I} num \, x_i) \wedge (\bigwedge_{j \in J} tree \, y_j)) \rightarrow (\exists ! \bar{z} \bigwedge_{k \in K} (tree \, z_k \wedge z_k = t_k(\bar{x}, \bar{y}, \bar{z})))$,
4   $\forall x \forall y \, x < y \rightarrow (num \, x \wedge num \, y)$,
5   $\forall x \forall y \, num \, x + y \leftrightarrow num \, x \wedge num \, y$,
6   $\forall x \, num \, -x \leftrightarrow num \, x$,
7   $\forall \bar{x} \, tree \, h\bar{x}$,
8   $\forall x \forall y \, (num \, x \wedge num \, y) \rightarrow x + y = y + x$,
9   $\forall x \forall y \forall z \, (num \, x \wedge num \, y \wedge num \, z) \rightarrow x + (y + z) = (x + y) + z$,
10 $\forall x \, num \, x \rightarrow x + 0 = x$,
11 $\forall x \, num \, x \rightarrow x + (-x) = 0$,
12_n $\forall x \, num \, x \rightarrow (nx = 0 \rightarrow x = 0)$,
13_n $\forall x \, num \, x \rightarrow \exists ! y \, num \, y \wedge ny = x$,
14 $\forall x \, num \, x \rightarrow \neg x < x$,
15 $\forall x \forall y \forall z \, num \, x \wedge num \, y \wedge num \, z \rightarrow ((x < y \wedge y < z) \rightarrow x < z)$,
16 $\forall x \forall y \, (num \, x \wedge num \, y) \rightarrow (x < y \vee x = y \vee y < x)$,
17 $\forall x \forall y \, (num \, x \wedge num \, y) \rightarrow (x < y \rightarrow (\exists z \, num \, z \wedge x < z \wedge z < y))$,
18 $\forall x \, num \, x \rightarrow (\exists y \, num \, y \wedge x < y)$,
19 $\forall x \, num \, x \rightarrow (\exists y \, num \, y \wedge y < x)$,
20 $\forall x \forall y \forall z \, (num \, x \wedge num \, y \wedge num \, z) \rightarrow (x < y \rightarrow (x + z < y + z))$,
21 $\forall x \, (\neg num \, x) \leftrightarrow tree \, x$
22 $0 < 1$,

where n is a non-null integer, f and g are two distinct function symbols taken from F, $h \in F - \{+, -, 0, 1\}$, x, y, z are variables, $\bar{x}$ is a vector of variables x_i, $\bar{y}$ is a vector of variables y_i, $\bar{z}$ is a vector of distinct variables z_i, I and J are finite possibly empty sets, and where $t_k(\bar{x}, \bar{y}, \bar{z})$ is a term which begins by a function symbol f_k element of $F - \{0, 1\}$ followed by variables taken from $\bar{x}$ or $\bar{y}$ or $\bar{z}$. Moreover, if $f_k \in \{+, -\}$ then $t_k(\bar{x}, \bar{y}, \bar{z})$ contains at least one variable taken from $\bar{y}$ or $\bar{z}$. The axiom 3 shows that all models of T contain infinite trees. In fact we have $T \models \exists! z \, z = fz \wedge tree \, z$ for $I = J = \emptyset$. In this case, the tree z is an infinite tree of the form $f(f(f(...)))$. Note that we have not $T \models \forall x num \, x \rightarrow (\exists! z \, z = x + x \wedge tree \, z)$, since we have $T \models num \, x \leftrightarrow num \, (x + x)$ according to axiom 5 which contradicts $tree \, z$ and $z = x + x$. This is why we have a condition if f_k belongs to $\{+, -\}$.

This theory has as model (possibly) infinite trees whose nodes are labelled by $Q \cup F$ such that each subtree labelled by $Q \cup \{+, -\}$ is evaluated in Q and reduced to a leaf labeled by an element of Q.

Let us now introduce an example of *constraints* in T. Let us consider the following two-player game: An ordered pair (n, m) of non-negative rational numbers is given and one after another each player subtracts 1 or 2 from n or m but keeping n and m non-negative. The first player who cannot play any more has lost.

Suppose that it is the turn of player A to play. A position (n, m) is called *k-winning* if, no matter the way the other player B plays, it is always possible for A to win, after having made at most k moves. The constraint expressing that a position x is k-winning is:

$$winning_k(x) \leftrightarrow \begin{bmatrix} \exists y \, move(x, y) \wedge \neg(\exists x \, move(y, x) \wedge \\ \neg(\exists y \, move(x, y) \wedge \neg(\exists x \, move(y, x) \wedge \neg(... \wedge \\ \underbrace{\neg(\exists y \, move(x, y) \wedge \neg(\exists x \, move(y, x) \wedge \neg(false\,)) ...)}_{2k} \end{bmatrix}$$

Each position (n, m) is represented by $c(i, j)$ with c a function symbol of arity 2 and $i, j \in Q$. The constraint $move(x, y)$ is defined by

$$\begin{bmatrix} (\exists i \exists j \, x = c(i, j) \wedge y = c(i - 1, j) \wedge i > 1 \wedge j > 0) \vee \\ (\exists i \exists j \, x = c(i, j) \wedge y = c(i - 2, j) \wedge i > 2 \wedge j > 0) \vee \\ (\exists i \exists j \, x = c(i, j) \wedge y = c(i, j - 1) \wedge i > 0 \wedge j > 1) \vee \\ (\exists i \exists j \, x = c(i, j) \wedge y = c(i, j - 2) \wedge i > 0 \wedge j > 2) \vee \\ (\neg(\exists i \exists j \, x = c(i, j) \wedge num \, i \wedge num \, j) \wedge x = y) \end{bmatrix}$$

By replacing the definition of *move* in the constraint $winning_k(x)$, we have a first-order constraint with one free variable x in the theory T of evaluated trees. Solving this constraint means finding all the positions x which are k-winning.

3   Block and Quantified Block in T

We will now present structured formulas called blocks and show some of their properties. Essentially a block is a conjunction of atomic formulas where all the variables are well typed and which gives enough informations to be locally solved. We will also define a mechanism to decompose each quantified block in three quantified blocks having interesting properties that will help us for solving first-order constraints on quantified blocks.

3.1 Basic Formula and Block in T

Suppose that the variables of V are ordered by a linear strict and dense order relation without endpoints, denoted by " $\succ$ ". For each formula φ, the bound variables are renamed such that for each sub-formula of φ we have $x \succ y$ for each bound variable x and each free variable y. We denote by $\Sigma_{i=1}^{n} t_i$ the term $\overline{t_1 + \ldots + t_n} + 0$ with $\overline{t_1 + \ldots + t_n}$ the term $t_1 + \ldots + t_n$ where all the terms 0 have been removed.

Let $f \in F - \{0, 1\}$, $a_0 \in Z$ and $a_i \in Z$. We call *leader* of the equation $x_0 = f x_1 \ldots x_n$ or $x_0 = x_1$ the variable x_0. We call *leader* of the formula $\Sigma_{i=1}^{n} a_i x_i = a_0 1$ the greatest variable x_k (in the order $\succ$) such that $a_k \neq 0$.

Let $f \in F$, $a_0 \in Z$ and $a_i \in Z$. We call *basic formula* every conjunction α of formulas of the form:

- *true*, *false*, *num x*, *tree x*,
- $x = y$, $x = f y_1 \ldots y_n$, $\Sigma_{i=1}^{n} a_i x_i = a_0 1$, $\Sigma_{i=1}^{n} a_i x_i < a_0 1$.

The formulas *num x* and *tree x* are called *typing constraints*. The formulas $x = y$, $x = f y_1 \ldots y_n$, $\Sigma_{i=1}^{n} a_i x_i = a_0 1$ are called *equations*. The formula $\Sigma_{i=1}^{n} a_i x_i < a_0 1$ is called *inequation*. Let α be a basic formula:

(1) We say that "*num x is a consequence of* α" iff α contains at least one of the following sub-formulas: *num x*, $x = y \wedge num\ y$, $y = x \wedge num\ y$, $x = -y \wedge num\ y$, $y = -x \wedge num\ y$, $z = y + x \wedge num\ z$, $z = x + y \wedge num\ z$, $x = y + z \wedge num\ z \wedge num\ y$, $x = 0$, $x = 1$, $\Sigma_i a_i x_i = a_0 1$ or $\Sigma_i a_i x_i < a_0 1$ and x is one of the x_i's.

(2) We say that "*tree x is a consequence of* α" iff α contains at least one of the following sub-formulas: *tree x*, $x = y \wedge tree\ y$, $y = x \wedge tree\ y$, $x = -y \wedge tree\ y$, $y = -x \wedge tree\ y$, $x = y + z \wedge tree\ z$, $x = z + y \wedge tree\ z$, $y = x + z \wedge tree\ y \wedge num\ z$, $y = z + x \wedge tree\ y \wedge num\ z$, $x = h y_1 \ldots y_n$, with $h \in F - \{+, -, 0, 1\}$.

(3) We call *tree-section* of α the conjunction α_t of the sub-formulas of α of the form:

- *true*, *tree x*,
- $x = y$ or $x = f y_1 \ldots y_n$, with $f \in F - \{0, 1\}$ and where x is such that *tree x* is a sub-formula of α.

This tree-section α_t is called *formatted* iff all the left-hand sides of the equations of α_t are distinct and for each equation $x = y$ of α_t we have $x \succ y$.

(4) We call *numeric-section* of α the conjunction α_n of sub-formulas of α of the form:

- *true*, *false*, $\Sigma_{i=1}^{n} a_i x_i = a_0 1$, $\Sigma_{i=1}^{n} a_i x_i < a_0 1$, *num x*,
- $x = y$, $x = -y$, $x = y + z$, where x is such that *num x* is a sub-formula of α.

This numeric-section α_n is called *consistent* iff $T \models \exists \bar{x} \, \alpha_n$ with $\bar{x} = var(\alpha_n)$ and *formatted* iff

- α_n does not contain sub-formulas of the form $x = y$, $x = -y$, $x = y + z$, $0 = a_0 1$, $0 < a_0 1$, with $a_0 \in \mathbf{Z}$
- α_n is consistent and each leader of the equations of α_n has one occurrence in only one the equations of α_n and no occurrence in the inequations of α_n.

(5) The variable u is called *reachable* in $\exists \bar{x} \alpha$ if u is a free variable in $\exists \bar{x} \alpha$ or α has a sub-formula of the form $y = t(u) \wedge tree\ y$ with $t(u)$ a term containing u and y a reachable variable. In the last case, the equation $y = t(u)$ is also called reachable in $\exists \bar{x} \alpha$.

Example: In the formula $\exists xyz\ w = fxy \wedge z = v \wedge tree\ w$, the variables w, v, x, y are reachable because w, v are free and x and y occur in the sub-formula $w = fxy \wedge tree\ w$. The variable z is not reachable and since z is bound and v is free, they must be such that $z \succ v$. The equation $w = fxy$ is reachable while the equation $z = v$ is not.

We call *block* every basic formulas α such that for each variable x in α either *num x* or *tree x* is a sub-formula of α and α does not contain sub-formulas of the form:

- $x = 0 \wedge tree\ x$, $x = 1 \wedge tree\ x$,
- $x = y \wedge num\ x \wedge tree\ y$, $x = y \wedge tree\ x \wedge num\ y$,
- $x = -y \wedge tree\ x \wedge num\ y$, $x = -y \wedge num\ x \wedge tree\ y$
- $x = y + z \wedge num\ x \wedge tree\ y$, $x = y + z \wedge num\ x \wedge tree\ z$, $x = h\bar{y} \wedge num\ x$,
- $x = y + z \wedge tree\ x \wedge num\ y \wedge num\ z$,
- $\Sigma_{i=1}^{n} a_i x_i = a_0 1 \wedge tree\ x_k$, $\Sigma_{i=1}^{n} a_i x_i < a_0 1 \wedge tree\ x_k$

with $h \in F - \{+, -, 0, 1\}$, $k \in \{1, ..., n\}$, $a_0 \in \mathbf{Z}$ and $a_i \in \mathbf{Z}$.

Since each variable x in a block is typed i.e. occurs in a sub-formula of the form *num x* or *tree x*, every block α can be divided into two disjoint sections: a tree-section and a numeric-section.

A block α without equations is called *relation block*. A block α without inequations and where each variable has an occurrence in at least one of the equations of α is called *equation* block. A block α is called *solved* iff its tree-section and numerical-section are formatted.

3.2   Decomposition of Quantified Solved Blocks

Let ψ be a formula. Let $\bar{x}$ be a vector of variables and α a solved block such that for all unreachable quantified variable u in $\exists \bar{x} \alpha$ and all reachable quantified variable v in $\exists \bar{x} \alpha$ we have $u \succ v$. We call *decomposition* of the formula $\exists \bar{x} \alpha \wedge \psi$ the formula

$$\exists \bar{x}^1 \alpha^1 \wedge (\exists \bar{x}^2 \alpha^2 \wedge (\exists \bar{x}^3 \alpha^3 \wedge \psi))), \tag{1}$$

obtained as follows : Let X be the set of the variables in $\bar{x}$. Let us decompose the set X into two disjoint subsets: X_r (the set of the elements of X which are reachable in $\exists \bar{x} \alpha$) and X_u. Let *Lead* be the set of the leaders of the equations of α. We have:

- $\bar{x}^1$ is the vector of the variables of X_r.
- $\bar{x}^2$ is the vector of the variables of $X_u - Lead$.

- $\bar{x}^3$ is the vector of the variables of $X_u \cap Lead$.
- α^1 is of the form $\alpha_1^1 \wedge \alpha_2^1$ where α_1^1 is the conjunction of all the equations in $\exists \bar{x}\alpha$ whose leader is reachable, α_2^1 is the conjunction of all the typing constraints of α which concern variables of $var(\alpha_1^1)$.
- α^2 is of the form $\alpha_1^2 \wedge \alpha_2^2$ where α_1^2 is the conjunction of all the inequations of α and α_2^2 is the conjunction of all the typing constraints of α which do not concern variables of $\bar{x}^3$.
- α^3 is of the form $\alpha_1^3 \wedge \alpha_2^3$ where α_1^3 is the conjunction of the other equations and α_2^3 is the conjunction of all the typing constraints of α which concern the variables of $var(\alpha_1^3)$. The restriction on the order $\succ$ of the quantified unreachable and reachable variables is due to an aim to get as leader of the equations of the numeric section of α unreachable variables. If one quantified leader is reachable then we deduce that all the quantified variables of this equation are reachable. This condition will help us for the algorithm of resolution given at Section 4. The intuitions behind this decomposition come from an aim to decompose a quantified solved block into three embedded sections each one having particular properties that enable us either to remove quantifiers or make special distributions in ψ and reduce the size of the formula $\exists \bar{x}\alpha \wedge \psi$.

Let A be the set of the solved blocks. Let A^1 be the set of the formulas of the form $\exists \bar{x}^1 \alpha^1$, where α^1 is a solved equation block and all the variables of $\bar{x}^1$ are reachable in $\exists \bar{x}^1 \alpha^1$. Let A^2 be the set of the solved relation blocks.

Property 3.2.1. *For all decomposed formula of the form (1) we have :* $\exists \bar{x}^1 \alpha^1 \in A^1$, $\alpha^2 \in A^2$, $\alpha^3 \in A$ *and* $T \models \forall \bar{x}^2 \alpha^2 \rightarrow \exists! \bar{x}^3 \alpha^3$.

Example 3.2.2. *Let* v, w, x, y, z *be variables such that* $w \succ y \succ z \succ x \succ v$. *Let us decompose the formula*

$$\exists wxyz \begin{bmatrix} v = fvx \wedge w + 2x + (-2)z = 1 \wedge y + 3z = 0 \wedge \\ z < 1 \wedge 3z + 2x < 0 \wedge \\ tree\ v \wedge num\ w \wedge num\ x \wedge num\ y \wedge num\ z \end{bmatrix} \tag{2}$$

The reachable variables in the formula (2) are v *and* x. *We have* $X_r = \{x, v\}$, $X_u = \{w, y, z\}$ *and* $Lead = \{v, w, y\}$. *Since* $w \succ y \succ z \succ x$ *then the formula (2) is equivalent in* T *to the decomposed formula*

$$\begin{bmatrix} \exists x\ v = fvx \wedge tree\ v \wedge num\ x \wedge \\ (\exists z\ z < 1 \wedge 3z + 2x < 0 \wedge num\ z \wedge numx \wedge tree\ v \wedge \\ (\exists wy\ w + 2x + (-2)z = 1 \wedge y + 3z = 0 \wedge num\ w \wedge num\ x \wedge num\ y \wedge num\ z)) \end{bmatrix}$$

Note that the elements of A^1 *does not accept elimination of quantifiers, this is due to the fact that all the variables of* $\bar{x}^1$ *are reachable in* $\exists \bar{x}^1 \alpha^1$. *Indeed in the formula* $\exists x\ v = fvx$ *the quantification* $\exists x$ *can not be eliminated in* T.

In all what follows we will use the notations $\bar{x}^1$, $\bar{x}^2$, $\bar{x}^3$, α^1, α^2, α^3 to refer to the decomposition of the formula $\exists \bar{x}\alpha$.

4 Solving First-Order Constraints in T

4.1 Working and General Solved Formulas

Definition 4.1.1. *A normalized formula φ of depth $d \geq 1$ is a formula of the form*

$$\neg(\exists \bar{x}\, \alpha \wedge \bigwedge_{i \in I} \varphi_i), \tag{3}$$

with I a finite (possibly empty) set, α a basic formula and the φ_i normalized formulas of depth d_i and $d = 1 + \max\{0, d_1, ..., d_n\}$.

Property 4.1.2. *Every formula is equivalent in T to a normalized formula.*

Definition 4.1.3. *A working formula is a normalized formula in which all the occurrences of $\neg$ are of the form $\neg^k$ with $k \in \{0, ..., 9\}$ and such that each occurrence of a sub-formula of the form*

$$\phi = \neg^k(\exists \bar{x}\, \alpha^c \wedge \alpha^p \wedge \bigwedge_{i \in I} \varphi_i), \tag{4}$$

*has $\alpha^p = true$ if $k = 0$ and **satisfies the first k conditions** of the following condition list if $k > 0$. Here α^p is a solved block and is called propagated constraint section, α^c is a basic formula and is called core constraint section, the φ_i are working formulas, and in the conditions: $\beta^p \wedge \beta^c$ is the conjunction of the equations and relations of the immediate top-working formula ψ of ϕ if it exists. i.e. $\psi = \neg^k(\exists \bar{y}\beta^c \wedge \beta^p \wedge \phi \wedge \bigwedge_{j \in J} \phi_j)$ where ϕ is the formula (4) and ϕ_j are any working formulas.*

1. *if ψ exists then $T \models \alpha^p \wedge \alpha^c \rightarrow \beta^p \wedge \beta^c$, and the tree-sections of α^p and $\beta^c \wedge \beta^p$ have the same set of left-hand side of equations,*
2. *the tree-section of $\alpha^p \wedge \alpha^c$ is formatted and the formula $\alpha^p \wedge \alpha^c$ does not contain tree $x \wedge$ num x for any variable x,*
3. *$\alpha^p \wedge \alpha^c$ is a block,*
4. *the numeric-section of $\alpha^p \wedge \alpha^c$ is consistent, and we have $u \succ v$ for u any unreachable variable in $\bar{x}$ and v any reachable variable in $\bar{x}$,*
5. *$\alpha^p \wedge \alpha^c$ is a solved block,*
6. *α^p is the formula $\beta^c \wedge \beta^p$ if ψ exists, and is the formula true otherwise. The formula α^c is a solved block and for each relation num x (or tree x) in α^p, if x does not occur in an equation or inequation of α^c then num x (resp. tree x) does not occur in α^c,*
7. *$(\exists \bar{x}\, \alpha^c)$ is decomposable into $(\exists \bar{x}^1\, \alpha^{c1} \wedge (\exists \bar{x}^2\, \alpha^{c2} \wedge (\exists \varepsilon\, true)))$,*
8. *$(\exists \bar{x}\, \alpha^c)$ is decomposable into $(\exists \bar{x}^1\, \alpha^{c1} \wedge (\exists \varepsilon\, \alpha^{c2} \wedge (\exists \varepsilon\, true)))$,*
9. *$(\exists \bar{x}\, \alpha^c)$ is decomposable into $(\exists \bar{x}^1\, \alpha^{c1} \wedge (\exists \varepsilon\, true \wedge (\exists \varepsilon\, true)))$.*

We use k in order to be able to control the execution of our rewriting rules on working formulas. We strongly insist on the fact that $\neg^k$ does not mean that the normalized formula satisfies only the kth condition but all the conditions i with $1 \leq i \leq k$. We call *initial* working formula a working formula of the form

$$\neg^6(\exists \varepsilon\, true \wedge \bigwedge_{i \in I} \varphi_i)$$

with φ_i working formulas where all negation symbols $\neg^k$ have $k = 0$ and all propagated constraint sections are reduced to *true*. We call *final* working formula a formula of the form

$$\neg^7(\exists\bar{\varepsilon}\ true \wedge \bigwedge_{i\in I} \neg^8(\exists\bar{x}_i\ \alpha_i^c \wedge \alpha_i^p \wedge \bigwedge_{j\in J_i} \neg^9(\exists\bar{y}_{ij}\ \beta_{ij}^c \wedge \beta_{ij}^p))), \qquad (5)$$

where all the β_{ij}^c are different from *true*.

Definition 4.1.4. *A* general solved formula *is a formula of the form*

$$\exists\bar{x}^1\ \alpha^1 \wedge \alpha^2 \wedge \bigwedge_{i\in I} \neg(\exists\bar{y}_i^1\ \beta_i^1), \qquad (6)$$

where $\exists\bar{x}^1\ \alpha^1 \in A^1$, $\alpha^2 \in A^2$, $\exists\bar{y}_i^1\ \beta_i^1 \in A^1$, *all the* $\alpha^1 \wedge \alpha^2 \wedge \beta_i^1$ *are solved blocks and all the* β_i^1 *are different from true.*

According to the properties of $\neg^8$ and $\neg^9$, in the final working formula (5), $\alpha_i^p = true$ and $\beta_{ij}^p = \alpha_i^p \wedge \alpha_i^c$. Thus the formula (5) is equivalent in T to the following disjunction of general solved formulas

$$\bigvee_{i\in I}(\exists\bar{x}_i\ \alpha_i^c \wedge \bigwedge_{j\in J_i} \neg(\exists\bar{y}_{ij}\ \beta_{ij}^c)) \qquad (7)$$

Property 4.1.5. *Let* φ *be a general solved formula of the form* (6). *If* φ *has no free variables then* φ *is the formula true, otherwise neither* $T \models \varphi$ *nor* $T \models \neg\varphi$ *and the solutions of the free variables of* φ *are explicit.*

This result is very important because it shows that for each solved formula φ containing at least one free variable there exists a set of solutions and a set of non-solutions, i.e. φ is neither true nor false in T. A similar result has been shown for the finite trees of J. Lassez [17] and the rational trees of M. Maher [20]. Note also that in all our proofs [8] we have not used the famous independence of inequations [4,16,6,18] but only the condition that the signature of T is infinite (F is infinite) which implies in this case the independence of the inequations.

4.2   Main Idea

The general algorithm for solving first-order constraints in T uses a system of rewriting rules. The main idea is to transform an initial working formula of depth d into a final working formula of depth less than or equal to three. The transformation is done in two steps:

(1) The first step is a top-down simplification and propagation. In each sub-working formula, $\alpha^c \wedge \alpha^p$ is transformed to a solved block, then $\exists\bar{x}\alpha^c$ is decomposed into three parts as in subsection 3.2. The third part is eliminated and added to the core-constraint section of the immediate sub-working formulas using a special property of the quantifier $\exists!$. The constraints of the two other parts

in α^p are propagated to the propagated-constraint section of the immediate sub-working formulas. In this step, the rules 1 to 24 are applied and transform the initial working formula into a working formula where each negation symbol is of the form $\neg^7$.

(2) The second step is a bottom-up simplification and elimination of quantifiers. This step is done by the rules 25 to 28. In each sub-working formula of depth one or two, the rule 25 eliminates quantified variables of the second part of the decomposition (the third one had been already removed in the first step). The rule 26 eliminates the constraints of the second part in the deepest level. Each sub-working formula of depth 3 is transformed step by step to a conjunction of working formulas of depth 2 by the rule 28 using a property of the quantifier $\exists$?. The transformations in this step can create new sub-working formulas where the first step needs to be done. At the end of the transformation, we obtain a final working formula of depth less than or equal to 3.

4.3   Rewriting Rules

We present in Figure 1 the rewriting rules which transform an initial working formula into an equivalent final working formula. To apply the rule $p_1 \implies p_2$ to the working formula p means to replace in p, a sub-formula p_1 by the formula p_2, by considering that the connector $\wedge$ is associative and commutative.

In all these rules, α is a basic formula, φ and ψ are conjunctions of working formulas.

In the rules 1 to 14, the equations and relations in α^c and α^p are mixed by considering the connector $\wedge$ associative and commutative. In these rules, except the rule 6, all modifications in the right hand side are done in α^c, since α^p is a solved block.

In the rule 2, f and g are two distinct function symbols taken from F. The rules 4, 6, 7, are applied only if $x \succ y$. This condition prevents infinite loops and makes the procedure terminating. In the rule 5, the equation $x = fz_1...z_n$ does not belong to α^p. In the rule 6, if the equation $x = fz_1...z_n$ belongs to α^p, then $x = y \wedge tree\ y$ is moved to α^p. In the rule 7, the equation $x = z$ does not belong to α^p.

In the rule 9, $a_0 > 0$. In the rules 13 and 14 the variable x_k is the leader of the equation $\Sigma_i a_i x_i = a_0 1$ and $b_k \neq 0$. Moreover, the equation $\Sigma_j b_j x_j = b_0 1$ does not belong to α^p. In the rule 14, the relation $\Sigma_j b_j x_j < b_0 1$ does not belong to α^p and $\lambda = 1$ if $a_k > 0$ and $\lambda = -1$ otherwise.

In the rule 15, the tree section of $\alpha^c \wedge \alpha^p$ is formatted and there is no subformula in $\alpha^c \wedge \alpha^p$ of the form $num\ x \wedge tree\ x$. In the rule 16 respectively 17, the typing constraint $num\ z$, respectively $tree\ z$ is not in $\alpha^c \wedge \alpha^p$ and is a consequence of $\alpha^c \wedge \alpha^p$. In the rule 18, z does not have typing constraints in $\alpha^c \wedge \alpha^p$ and neither $num\ z$ nor $tree\ z$ is a consequence of $\alpha^c \wedge \alpha^p$.

In the rule 19, $\alpha^c \wedge \alpha^p$ is a block. In the rule 20, the numeric section of $\alpha^c \wedge \alpha^p$ is inconsistent. In the rule 21, the unreachable variables in $\bar{x}$ are renamed if necessary such that $u \succ v$ for each unreachable variable u and each reachable variable v in $\bar{x}$ and the numeric section of $\alpha^c \wedge \alpha^p$ is consistent. The consistency

$1 \quad \neg^1(\exists \bar{u} \, num \, x \wedge tree \, x \wedge \alpha \wedge \varphi) \qquad \Longrightarrow true$

$2 \quad \neg^1(\exists \bar{u} \, x = f\bar{y} \wedge x = g\bar{z} \wedge tree \, x \wedge \alpha \wedge \varphi) \Longrightarrow true$

$3 \quad \neg^1(\exists \bar{u} \, x = x \wedge \alpha \wedge \varphi) \qquad\qquad \Longrightarrow \neg^1(\exists \bar{u} \, \alpha \wedge \varphi)$

$4 \quad \neg^1(\exists \bar{u} \, y = x \wedge tree \, x \wedge \alpha \wedge \varphi) \qquad \Longrightarrow \neg^1(\exists \bar{u} \, x = y \wedge tree \, x \wedge \alpha \wedge \varphi)$

$5 \quad \neg^1 \begin{bmatrix} \exists \bar{u} \, x = fy_1...y_n \wedge x = fz_1...z_n \wedge \\ tree \, x \wedge \alpha \wedge \varphi \end{bmatrix} \Longrightarrow \neg^1 \begin{bmatrix} \exists \bar{u} \, x = fy_1...y_n \wedge \bigwedge_i y_i = z_i \wedge \\ tree \, x \wedge \alpha \wedge \varphi \end{bmatrix}$

$6 \quad \neg^1 \begin{bmatrix} \exists \bar{u} \, x = y \wedge x = fz_1...z_n \wedge \\ tree \, x \wedge tree \, y \wedge \alpha \wedge \varphi \end{bmatrix} \Longrightarrow \neg^1 \begin{bmatrix} \exists \bar{u} \, x = y \wedge y = fz_1...z_n \wedge \\ tree \, x \wedge tree \, y \wedge \alpha \wedge \varphi \end{bmatrix}$

$7 \quad \neg^1(\exists \bar{u} \, x = y \wedge x = z \wedge tree \, x \wedge \alpha \wedge \varphi) \Longrightarrow \neg^1(\exists \bar{u} \, x = y \wedge y = z \wedge tree \, x \wedge \alpha \wedge \varphi)$

$8 \quad \neg^4(\exists \bar{u} \, 0 = 0 \wedge \alpha \wedge \varphi) \qquad\qquad \Longrightarrow \neg^4(\exists \bar{u} \, \alpha \wedge \varphi)$

$9 \quad \neg^4(\exists \bar{u} \, 0 < a_0 1 \wedge \alpha \wedge \varphi) \qquad\quad \Longrightarrow \neg^4(\exists \bar{u} \, \alpha \wedge \varphi)$

$10 \; \neg^4 \begin{bmatrix} \exists \bar{u} \, x = y \wedge \\ num \, x \wedge num \, y \wedge \alpha \wedge \varphi \end{bmatrix} \Longrightarrow \neg^4 \begin{bmatrix} \exists \bar{u} \, x + (-1y) = 0 \wedge \\ num \, x \wedge num \, y \wedge \alpha \wedge \varphi \end{bmatrix}$

$11 \; \neg^4 \begin{bmatrix} \exists \bar{u} \, x = -y \wedge \\ num \, x \wedge num \, y \wedge \alpha \wedge \varphi \end{bmatrix} \Longrightarrow \neg^4 \begin{bmatrix} \exists \bar{u} \, x + y = 0 \wedge \\ num \, x \wedge num \, y \wedge \alpha \wedge \varphi \end{bmatrix}$

$12 \; \neg^4 \begin{bmatrix} \exists \bar{u} \, x = y + z \wedge num \, x \wedge \\ num \, y \wedge num \, z \wedge \alpha \wedge \varphi \end{bmatrix} \Longrightarrow \neg^4 \begin{bmatrix} \exists \bar{u} \, x + (-1y) + (-1z) = 0 \wedge \\ num \, x \wedge num \, y \wedge num \, z \wedge \alpha \wedge \varphi \end{bmatrix}$

$13 \; \neg^4 \begin{bmatrix} \exists \bar{u} \, \Sigma_{i=1}^n a_i x_i = a_0 1 \wedge \\ \Sigma_{i=1}^n b_i x_i = b_0 1 \wedge \\ \alpha \wedge \varphi \end{bmatrix} \Longrightarrow \neg^4 \begin{bmatrix} \exists \bar{u} \, \Sigma_{i=1}^n a_i x_i = a_0 1 \wedge \\ \Sigma_{i=1}^n (b_k a_i - a_k b_i) x_i = (b_k a_0 - a_k b_0) 1 \wedge \\ \alpha \wedge \varphi \end{bmatrix}$

$14 \; \neg^4 \begin{bmatrix} \exists \bar{u} \, \Sigma_{i=1}^n a_i x_i = a_0 1 \wedge \\ \Sigma_{i=1}^n b_i x_i < b_0 1 \wedge \\ \alpha \wedge \varphi \end{bmatrix} \Longrightarrow \neg^4 \begin{bmatrix} \exists \bar{u} \, \Sigma_{i=1}^n a_i x_i = a_0 1 \wedge \\ \Sigma_{i=1}^n \lambda(b_k a_i - a_k b_i) x_i < (b_k a_0 - a_k b_0) 1 \wedge \\ \alpha \wedge \varphi \end{bmatrix}$

$15 \; \neg^1(\exists \bar{x} \, \alpha^c \wedge \alpha^p \wedge \varphi) \qquad\qquad \Longrightarrow \neg^2(\exists \bar{x} \, \alpha^c \wedge \alpha^p \wedge \varphi)$

$16 \; \neg^2(\exists \bar{x} \, \alpha^c \wedge \alpha^p \wedge \varphi) \qquad\qquad \Longrightarrow \neg^1(\exists \bar{x} \, num \, z \wedge \alpha^c \wedge \alpha^p \wedge \varphi)$

$17 \; \neg^2(\exists \bar{x} \, \alpha^c \wedge \alpha^p \wedge \varphi) \qquad\qquad \Longrightarrow \neg^1(\exists \bar{x} \, tree \, z \wedge \alpha^c \wedge \alpha^p \wedge \varphi)$

$18 \; \neg^2(\exists \bar{x} \, \alpha^c \wedge \alpha^p \wedge \varphi) \qquad\qquad \Longrightarrow \begin{bmatrix} \neg^1(\exists \bar{x} \, num \, z \wedge \alpha^c \wedge \alpha^p \wedge \varphi) \wedge \\ \neg^1(\exists \bar{x} \, tree \, z \wedge \alpha^c \wedge \alpha^p \wedge \varphi) \end{bmatrix}$

$19 \; \neg^2(\exists \bar{x} \, \alpha^c \wedge \alpha^p \wedge \varphi) \qquad\qquad \Longrightarrow \neg^3(\exists \bar{x} \, \alpha^c \wedge \alpha^p \wedge \varphi)$

$20 \; \neg^3(\exists \bar{x} \, \alpha^c \wedge \alpha^p \wedge \varphi) \qquad\qquad \Longrightarrow true$

$21 \; \neg^3(\exists \bar{x} \, \alpha^c \wedge \alpha^p \wedge \varphi) \qquad\qquad \Longrightarrow \neg^4(\exists \bar{x} \, \alpha^c \wedge \alpha^p \wedge \varphi)$

$22 \; \neg^4(\exists \bar{x} \, \alpha^c \wedge \alpha^p \wedge \varphi) \qquad\qquad \Longrightarrow \neg^5(\exists \bar{x} \, \alpha^c \wedge \alpha^p \wedge \varphi)$

$23 \; \neg^7 \begin{bmatrix} \exists \bar{x} \, \alpha^c \wedge \alpha^p \wedge \varphi \wedge \\ \neg^5(\exists \bar{y} \, \beta^c \wedge \beta^p \wedge \psi) \end{bmatrix} \Longrightarrow \neg^7 \begin{bmatrix} \exists \bar{x} \, \alpha^c \wedge \alpha^p \wedge \varphi \wedge \\ \neg^6(\exists \bar{y} \, \gamma^c \wedge \gamma^p \wedge \psi) \end{bmatrix}$

$24 \; \neg^6 \begin{bmatrix} \exists \bar{x} \, \alpha^c \wedge \alpha^p \wedge \\ \bigwedge_i \neg^0(\exists \bar{y}_i \beta_i^c \wedge \beta_i^p \wedge \varphi_i) \end{bmatrix} \Longrightarrow \neg^7 \begin{bmatrix} \exists \bar{x}^1 \bar{x}^2 \, \alpha^{c1} \wedge \alpha^{c2} \wedge \alpha^p \wedge \\ \bigwedge_i \neg^1(\exists \bar{y}_i \bar{x}^3 \gamma_i^c \wedge \gamma_i^p \wedge \varphi_i) \end{bmatrix}$

$25 \; \neg^7 \begin{bmatrix} \exists \bar{x} \, \alpha^c \wedge \alpha^p \wedge \\ \bigwedge_{i \in I} \neg^9(\exists \bar{y}_i \beta_i^c \wedge \beta_i^p) \end{bmatrix} \Longrightarrow \neg^8 \begin{bmatrix} \exists \bar{x}^1 \alpha^{c1} \wedge \alpha^{c2*} \wedge \alpha^p \wedge \\ \bigwedge_{i \in I'} \neg^9(\exists \bar{y}_i \beta_i^c \wedge \beta_i^{p*}) \end{bmatrix}$

$26 \; \neg^7 \begin{bmatrix} \exists \bar{x} \, \alpha^c \wedge \alpha^p \wedge \varphi \wedge \\ \neg^8(\exists \bar{y} \beta^c \wedge \beta^p) \end{bmatrix} \Longrightarrow \begin{bmatrix} \neg^7(\exists \bar{x} \, \alpha^c \wedge \alpha^p \wedge \varphi \wedge \neg^9(\exists \bar{y} \beta^{c1} \wedge \beta^p)) \wedge \\ \bigwedge_{i \in I} \neg^1(\exists \bar{x} \bar{y} \beta^p \wedge \beta^{c1} \wedge \beta_i^{c2*} \wedge \varphi_0) \end{bmatrix}$

$27 \; \neg^7 \begin{bmatrix} \exists \bar{x} \, \alpha^c \wedge \alpha^p \wedge \varphi \wedge \\ \neg^9(\exists \varepsilon true \wedge \beta^p) \end{bmatrix} \Longrightarrow true$

$28 \; \neg^7 \begin{bmatrix} \exists \bar{x} \, \alpha^c \wedge \alpha^p \wedge \varphi \wedge \\ \neg^8 \begin{bmatrix} \exists \bar{y} \, \beta^c \wedge \beta^p \wedge \\ \bigwedge_{i \in I} \neg^9(\exists \bar{z}_i \gamma_i^c \wedge \gamma_i^p) \end{bmatrix} \end{bmatrix} \Longrightarrow \begin{bmatrix} \neg^7(\exists \bar{x} \, \alpha^c \wedge \alpha^p \wedge \varphi \wedge \neg^8(\exists \bar{y} \beta^c \wedge \beta^p)) \wedge \\ \bigwedge_{i \in I} \neg^6(\exists \bar{x} \bar{y} \bar{z}_i \delta_i^c \wedge \delta_i^p \wedge \varphi_0) \end{bmatrix}$

Fig. 1. The rewriting rules

can be verified for example by using the first step of the Simplex. In the rule 22, $\alpha^c \wedge \alpha^p$ is a solved block.

In the rule 23, γ^c is obtained from β^c as follows: for every variable $x \in \text{var}(\beta^c)$, we add all the relations $num\ x$ or $tree\ x$ which are in β^p but not in β^c, and for all the variables y which do not occur in an equation or inequation of β^c we remove all relations $num\ y$ or $tree\ y$ which are both in β^c and β^p. The formula γ^p is the formula $\alpha^p \wedge \alpha^c$.

In the rule 24, $\exists \bar{x} \alpha^c$ is decomposed into $\exists \bar{x}^1 \alpha^{c1} \wedge (\exists \bar{x}^2 \alpha^{c2} \wedge (\exists \bar{x}^3 \alpha^{c3}))$, $\gamma_i^c = \beta_i^c \wedge \alpha^{c3}$ and $\gamma_i^p = \beta_i^p \wedge \alpha^{c1} \wedge \alpha^{c2} \wedge \alpha^p$.

The four rules 25, 26, 27 and 28 cannot be applied on the occurrence of $\neg^7$ of the first level of the general working formula. In the rule 25, all the $\beta_c i$ are different from $true$, I' is the set of $i \in I$ such that β_i^c does not contain occurrences of any variables in $\bar{x}^2$. The formula α^{c2*} is such that $T \models (\exists \bar{x}^2 \alpha^{c2}) \leftrightarrow \alpha^{c2*}$ and is computed using the Fourier quantifier elimination. The propagated-constraint section $\beta_i^{p*} = \alpha^{c1} \wedge \alpha^{c2*} \wedge \alpha^p$.

In the rule 26, φ is such that every negation symbol $\neg^k$ has $k \geq 6$, φ_0 is obtained from φ by replacing all occurrences of $\neg^k$ by $\neg^0$ and all propagated-constraint sections by $true$. Let β^2 be the formula obtained from β^{c2} by removing the multiple occurrences of typing constraints and for all the variables y which do not occur in an inequation of β^{c2} we remove all relations $num\ y$ or $tree\ y$ which are both in β^{c1} and β^{c2}. If β^2 is the formula $true$ then $I = \emptyset$, otherwise the β_i^{c2*} with $i \in I$ are obtained from β^2 as follows: Since $\beta^2 \in A^2$ then it is of the form

$$\left[\begin{array}{c} (\bigwedge_{\ell \in L} num\ z_\ell) \wedge (\bigwedge_{k \in K} tree\ v_k) \wedge \\ ((\bigwedge_{j \in J} \sum_{i=1}^{n} a_{ij} x_i < a_{0j}) \wedge \bigwedge_{m=1}^{n} num\ x_m) \end{array} \right],$$

thus $\neg \beta^2$ is of the form

$$\left[\begin{array}{c} (\bigvee_{\ell \in L} tree\ z_\ell) \vee (\bigvee_{k \in K} num\ v_k) \vee (\bigvee_{m=1}^{n} tree\ x_m) \vee \\ \bigvee_{j \in J} ((\sum_{i=1}^{n} a_{ij} x_i = a_{0j} 1 \wedge \bigwedge_{m=1}^{n} num\ x_m) \vee \\ (\sum_{i=1}^{n} (-a_{ij}) x_i < (-a_{0j}) 1 \wedge \bigwedge_{m=1}^{n} num\ x_m)) \end{array} \right]$$

Each element of this disjunction is a block and represents a formula β_i^{c2*}. Of course we have $T \models (\neg \beta^2) \leftrightarrow \bigvee_i \beta_i^{c2*}$.

In the rule 28, $I \neq \emptyset$, φ is such that every negation symbol $\neg^k$ has $k \geq 6$, φ_0 is obtained from φ by replacing all occurrences of $\neg^k$ by $\neg^0$ and all propagated-constraint sections by $true$. Moreover $\delta_i^p = \alpha^p$ and $\delta_i^c = \gamma_i^c \wedge \beta^c \wedge \alpha^c$.

Property 4.3.1. *Every repeated application of the precedent rewriting rules on an inital working formula terminates and produces an equivalent final working formula which does not contain new free variables.*

Corollary 4.3.2. *Every formula is equivalent in T either to true or to false or to a disjunction of general solved formulas, having at least one free variable, being equivalent neither to true nor to false in T and where the solutions of the free variables are expressed in a clear and explicit way.*

In fact, solving a constraint φ in T proceeds as follows:

1. Transform φ into a normalized formula, then into an initial working formula ϕ, which is equivalent to φ in T.
2. Transform ϕ into a final working formula ψ using the rewriting rules defined in the subsection 4.3.
3. Extract from ψ the equivalent disjunction of general solved formulas. If this disjunction contains the general solved formula $true$, then it is reduced to $true$.

Example: Let φ be the following constraint having i, j as free variables:

$$\exists x\, x = fij \wedge i > 0 \wedge tree\, x \wedge num\, i \wedge num\, j \wedge \neg(\exists k\, j = 2k \wedge num\, k).$$

We can see that $num\, j \wedge \neg(\exists k\, j = 2k \wedge num\, k)$ is always false in T since for every variable j, there exists a unique variable k such that $j = 2k$ (axiom 13_n). Let us transform φ into an initial working formula (the propagated-constraint sections are underlined):

$$\neg^6\neg^0(\exists x\, x = fij \wedge i > 0 \wedge tree\, x \wedge num\, j \wedge \underline{true} \wedge \neg^0(\exists k\, j = 2k \wedge num\, k \wedge \underline{true}))$$

After having applied the rules 24, 15, 16, 15, 19, 21, 22, 23 in this order, we obtain:

$$\neg^7\neg^6(\exists x\, x = fij \wedge i > 0 \wedge tree\, x \wedge num\, i \wedge num\, j \wedge \underline{true} \wedge \neg^0(\exists k\, j = 2k \wedge num\, k \wedge \underline{true}))$$

The rule 24 being applied changes the formula to:

$$\neg^7\neg^7 \left[\begin{matrix} i > 0 \wedge num\, i \wedge num\, j \wedge \underline{true}\wedge \\ \neg^1 \left[\begin{matrix} \exists xk\, x = fij \wedge j = 2k \wedge num\, k \wedge tree\, x\wedge \\ \underline{i > 0 \wedge num\, i \wedge num\, j} \end{matrix} \right] \end{matrix} \right]$$

After having applied on the sub-working formula $\neg^1(...)$ the rule 15, 19, 21, 12, 22, 23

$$\neg^7\neg^7 \left[\begin{matrix} i > 0 \wedge num\, i \wedge num\, j \wedge \underline{true}\wedge \\ \neg^6 \left[\begin{matrix} \exists xk\, x = fij \wedge j - 2k = 0 \wedge num\, k \wedge tree\, x\wedge \\ \underline{i > 0 \wedge num\, i \wedge num\, j} \end{matrix} \right] \end{matrix} \right]$$

The rule 24 is applied then we obtain:

$$\neg^7\neg^7(i > 0 \wedge num\, i \wedge num\, j \wedge \underline{true} \wedge \neg^7(\underline{true} \wedge \underline{i > 0 \wedge num\, i \wedge num\, j}))$$

The rules 25, 26 are applied in this order, giving:

$$\neg^7\neg^7(i > 0 \wedge num\, i \wedge num\, j \wedge \underline{true} \wedge \neg^9(\underline{true} \wedge \underline{i > 0 \wedge num\, i \wedge num\, j}))$$

Finally, by application of the rule 27, we obtain the final working formula $\neg^7 true$, which is equivalent to the empty disjunction of general solved formulas, i.e. $false$. Thus, the initial constraint φ is false in T and does not depend on the values of its free variables i and j. A such phenomena is impossible to detect using a decision procedure instead of a first-order constraint solver.

5   Conclusion

Quantified formulas over trees and rational numbers provide an expressive constraint language that is essential in applications such as program analysis and model checking. We have presented in this paper a first-order constraint solver in the theory of the evaluated trees. The algorithm is given in the form of 28 rewriting rules and its correctness implies the completeness of a theory built on the model of Prolog III. Our aim in this work was not only to decide proposition i.e. to decide if a formula without free variables is true or false in T but to express the solutions of any first-order constraint having free variables in a clear and explicit way.

S. Vorobyov [25] has shown that the problem of deciding if a proposition is true or not in the theory of trees is non-elementary, i.e. the complexity of all algorithms which solve it cannot be bound by a tower of powers of $2's$ (with a top down evaluation) with a fixed height. Thus, our algorithm must not escape this kind of complexity in the worst case. This is why we have used two strategies in the algorithm: a top down propagation of constraints and a bottom-up elimination of quantifiers and distribution. This technique can quickly detect (using propagation and local solving) sub-formulas which are equivalent to false and prevents us from solving a big working formula (i.e. a working formula of huge depth) which contradicts its top-working formula. We have recently programmed a similar algorithm only on the theory of finite or infinite trees and in spite of the high complexity we can solve formulas on two partners games involving 160 nested quantifiers [10].

Currently, we are trying to find other classes of theories T_i such that we can apply a similar technique to solve first-order constraints in the hybrid theories $T_i + Trees$. We are also working on a possibly CHR (Constraint Handling Rules) implementation of our solver.

Acknowledgements. We thank Alain Colmerauer for our many discussions and his help in this work. We dedicate to him this paper.

References

1. Benhamou, F., Van Caneghem, M.C.: Le manuel de Prolog IV, PrologIA, France (1996)
2. Bürckert, H.: Solving disequations in equational theories. In: Lusk, E.'., Overbeek, R. (eds.) 9th International Conference on Automated Deduction. LNCS, vol. 310, pp. 517–526. Springer, Heidelberg (1988)
3. Colmerauer, A.: An introduction to Prolog III. Communication of the ACM 33(7), 68–90 (1990)
4. Colmerauer, A.: Equations and disequations on finite and infinite trees. In: Proc of the 5th conf on generation of computer systems Tokyo, 1984, pp. 85–99 (1984)
5. Colmerauer, A.: Prolog and infinite trees. In: Clark, K.L., Tarnlund, S-A. (eds.) Logic Programming, pp. 231–251. Academic Press, London (1982)

6. Comon, H.: Résolution de contraintes dans des algèbres de termes. Rapport d'Habilitation, Université de Paris Sud (1992)
7. Courcelle, B.: Fundamental Properties of Infinite Trees. TCS 25(2), 95–169 (1983)
8. Dao, T., Djelloul, K.: Solving first-order constraints in evaluated trees. Technical report, Laboratoire d'informatique fondamentale d'Orlans, RR-2006-05, Full version of this paper with full proofs, http://www.univ-orleans.fr/lifo/rapports.php
9. Djelloul, K.: About the combination of trees and rational numbers in a complete first-order theory. In: Gramlich, B. (ed.) Frontiers of Combining Systems. LNCS (LNAI), vol. 3717, pp. 106–122. Springer, Heidelberg (2005)
10. Djelloul, K., Dao, T.: Solving First-Order formulas in the Theory of Finite or Infinite Trees. In: Proceeding of the 21st ACM Symposium on Applied Computing (SAC'06), pp. 7–14. ACM Press, New York (2006)
11. Djelloul, K., Dao, T.: Extension into trees of first order theories. In: Calmet, J., Ida, T., Wang, D. (eds.) AISC 2006. LNCS (LNAI), vol. 4120, pp. 53–67. Springer, Heidelberg (2006)
12. Djelloul, K.: Decomposable Theories. Journal of Theory and practice of Logic Programming (to appear, 2006)
13. Huet, G.: Resolution d'equations dans les langages d'ordre 1, 2,...ω. These d'Etat, Universite Paris 7. France (1976)
14. Jaffar, J.: Efficient unification over infinite terms. New Generation Computing 2(3), 207–219 (1984)
15. Jouannaud, J.P., Kirchner, C.: Solving Equations in Abstract Algebras: A Rule-Based Survey of Unification. Computational Logic - Essays in Honor of Alan Robinson, pp. 257–321. MIT Press, Cambridge (1991)
16. Lassez, J., Maher, M., Marriott, K.: Unification revisited. In: proceedings of the workshop on the foundations of deductive database and logic programming, pp. 587–625 (1986)
17. Lassez, J., Marriott, K.: Explicit representation of terms defined by counter examples. Journal of automated reasonning. 3, 301–317 (1987)
18. Lassez, J., McAloon, K.: Independence of negative constraints. In: Díaz, J., Orejas, F. (eds.) CAAP 1989 and TAPSOFT 1989. LNCS, vol. 351, pp. 19–27. Springer, Heidelberg (1989)
19. Maher, M.: Complete axiomatization of the algebra of finite, rational and infinite trees. Technical report, IBM - T.J.Watson Research Center (1988)
20. Maher, M., Stuckey, P.: On inductive inference of cyclic structures. Annals of mathematics and artificial intelligence 15(2), 167–208 (1995)
21. Martelli, A., Montanari, U.: An efficient unification algorithm. ACM Trans. on Languages and Systems 4(2), 258–282 (1982)
22. Paterson, M., Wegman, N.: Linear unification. Journal of Computer and Systems Science 16, 158–167 (1978)
23. Ramachandran, V., Van Hentenryck, P.: Incremental algorithms for constraint solving and entailment over rational trees. In: Shyamasundar, R.K. (ed.) Foundations of Software Technology and Theoretical Computer Science. LNCS, vol. 761, pp. 205–217. Springer, Heidelberg (1993)
24. Robinson, J.A.: A machine-oriented logic based on the resolution principle. JACM 12(1), 23–41 (1965)
25. Vorobyov, S.: An Improved Lower Bound for the Elementary Theories of Trees. In: McRobbie, M.A., Slaney, J.K. (eds.) Automated Deduction (CADE'96). LNCS, vol. 1104, pp. 275–287. Springer, Heidelberg (1996)

Extracting Microstructure in Binary Constraint Networks

Chavalit Likitvivatanavong and Roland H.C. Yap

School of Computing, National University of Singapore, Singapore

Abstract. We present algorithms that perform the extraction of partial assignments from binary Constraint Satisfaction Problems without introducing new constraints. They are based on a new perspective on domain values: we view a value not as a single, indivisible unit, but as a combination of value fragments. Applications include removing nogoods while maintaining constraint arity, learning nogoods in the constraint network, enforcing on neighborhood inverse consistency and removal of unsolvable sub-problems from the constraint network.

1 Introduction

Constraint Satisfaction Problems (CSP) is one of the most important modeling tools in AI with far-reaching applications. There are many variations on the standard CSP, such as weighted constraints, partial, or distributed CSP, but they are all based on the same notion: assigning values to variables in order to satisfy constraints among them.

Normally, values that can be assigned to a variable are treated as being inherently different from each other. In [1] however, a new viewpoint on domain values is proposed which takes into account the microstructure (a low-level graphical representation of a CSP based on compatibility of values). Basically, a value is composed of *label* and *support structure* and can be broken into several values as long as they correspond to the original one. Solutions involving values with the same label are indistinguishable from one another given the rest of the solutions being the same.

This definition of a domain value has been shown effective in reducing interchangeability within a variable domain by recombining value fragments to eliminate identical parts [1]. This paper also uses the microstructure but differs in that we only break values apart and do not form new values from the value fragments. Furthermore, it deals with a partition of the microstructure that spans many connected variables, as opposed to a single variable in [1].

We illustrate the main idea and one of its applications in Figure 1(i). There are three variables and eight values, and each line denotes a compatible pair of values. Suppose we want to designate the tuple (b, d, g) as a nogood (a partial assignment that cannot be extended to a full solution) the usual approach would be to add a new 3-ary constraint involving the three variables. It is unsafe to directly delete one of b, d, or g since solutions involving them may be inadvertently eliminated in the process.

F. Azevedo et al. (Eds.): CSCLP 2006, LNAI 4651, pp. 124–138, 2007.
© Springer-Verlag Berlin Heidelberg 2007

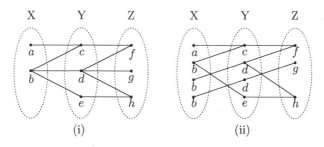

Fig. 1. Marking (b, d, g) in (i) as a nogood by eliminating one of b, d, or g, could remove some solutions, while doing so in (ii), an equivalent network, is completely safe

Figure 1(ii) depicts an equivalent CSP in which values are split into fragments as shown. Values having the same label (e.g. the three values with label "b") are differentiated by their structures. Solutions of (i) are the same as those of (ii) with respect to their support labels. The tuple (b, d, g) is isolated (by the process we will later describe) and thus we can simply delete one of the values involved to mark it as a nogood; the rest would be removed by simple arc consistency processing. Should this network be part of a larger one, the links to the external part would need to be modified so that they connect to the newly-created values instead (e.g a single link to b in (i) corresponds to three links to the three b in (ii)).

Our contribution includes a novel process of extracting partial assignments without introducing new constraints, along with the following applications:

- *Maintaining constraint arity.* Because no new constraint is introduced, the original constraint arity would remain unchanged. Algorithms that work only for binary CSPs [2] for instance would be usable even after higher-order nogoods are incorporated.
- *Learning nogoods by pruning.* Nogoods can be extracted and pruned from the microstructure, as shown in Figure 1(ii), rather than recorded [3]. This gives an alternative approach to deal with the problem of increasing memory requirements for recording nogoods.
- *Enforcing Neighborhood Inverse Consistency in one pass.* Like many other consistency algorithms, NIC [4] requires more than a single pass of propagation to fully achieve consistency. We show how to do it in a single pass.
- *Extracting unsolvable sub-problems.* The common approach taken is to decompose a CSP into several sub-problems [5]. Our method is more efficient since only the target sub-problem is extracted while everything else remains intact in the original CSP.

The paper is organized as follows. In Section 2 we cover the CSP preliminary and recall the formalism on domain values. In Section 3 we provide two algorithms, one for extracting simple nogoods and the other for extracting complex ones. along with their proofs of correctness. In Section 4, the process is generalized so that more complex microstructures can be separated. Applications are described in Section 5. We conclude in Section 6.

2   Background

Definition 1 (Binary Constraint Network). *A binary constraint network $\mathcal{P}$ is a triplet $(\mathcal{V}, \mathcal{D}, \mathcal{C})$ where $\mathcal{V}$ is a finite set of variables, $\mathcal{D} = \bigcup_{V \in \mathcal{V}} D_V$ where D_V is a finite set of possible values for V, and $\mathcal{C}$ is a finite set of constraints such that each $C_{XY} \in \mathcal{C}$ is a subset of $D_X \times D_Y$ indicating the possible pairs of values for X and Y, where $X \neq Y$. Without loss of generality, we assume that two values in different domains are distinct (i.e. if $a \in D_X$ then $a \notin D_Y$ for all $Y \neq X$). If $a \in D_X$ then we use var(a) to denote X. If $C_{XY} \in \mathcal{C}$, then $C_{YX} = \{(y,x) \mid (x,y) \in C_{XY}\}$ is also in $\mathcal{C}$.*

A binary constraint network is k-consistent if given $k - 1$ variables and $k - 1$ values that satisfy the constraints on these variables, we are able to find a value (called a support*) for any kth variable such that all the constraints on the k variables will be satisfied by the k values taken together. When $k = 2$ and 3 it is called Arc Consistency (AC) and Path Consistency (PC) respectively.*

The neighborhood *of variable V (N_V) is the set $\{W \mid C_{VW} \in \mathcal{C}\}$. We use n, e, and d to denote the number of variables, the number of constraints, and the maximum domain size. An* assignment *is a function $\pi : \mathcal{W} \subseteq \mathcal{V} \to \mathcal{D}$ such that $\pi(W) \in D_W$ for all $W \in \mathcal{W}$; we denote the function domain by $\mathrm{dom}(\pi)$. π is* consistent *if and only if for all $C_{XY} \in \mathcal{C}$ such that $X, Y \in \mathrm{dom}(\pi)$, $(\pi(X), \pi(Y)) \in C_{XY}$. π is a* solution *if and only if $\mathrm{dom}(\pi) = \mathcal{V}$ and π is consistent. We use $\mathcal{P} \mid_{\mathcal{W}}$ to denote the binary constraint network induced by $\mathcal{W} \subseteq \mathcal{V}$. A* Constraint Satisfaction Problem *involves finding one or more solutions to an associated constraint network or declaring it unsatisfiable.*

We extend the usual definition of a value to a 2-tuple in order to clearly distinguish between the syntax (value of the label) and the semantics (the support corresponding to the label).

Definition 2 (Values). *A value $a \in D_V$ is a 2-tuple (L, σ) where L is a set of labels, while σ, called* support structure*, is a function $\sigma : N_V \to 2^{\mathcal{D}}$ such that $\sigma(W) \subseteq D_W$ for any $W \in N_V$. We use L_a to denote the set of labels of value a and σ_a denotes the support structure of a. A value a must be* valid*, that is, $\sigma_a(W) = \{b \in D_W \mid (a, b) \in C_{VW}\}$ for any $W \in N_V$. The* local solutions *of a is the set $\{(s_1, \ldots, s_{|N_V|}) \in D_{W_1} \times D_{W_2} \times \ldots \times D_{W_{|N_V|}} \mid s_i \in \sigma_a(W_i), W_i \in N_V\}$, the enumeration of a's supports. The* size *of a (size(a)) is $\prod_{W \in N_V} |\sigma_a(W)|$, which equals the number of local solutions of a.*

Let $\mathcal{L} = \bigcup_{a \in \mathcal{D}} L_a$ be the set of all labels. A label-assignment *is a function $\lambda : \mathcal{W} \subseteq \mathcal{V} \to \mathcal{L}$ such that $\lambda(W) \in \bigcup_{a \in D_W} L_a$ for all $W \in \mathcal{W}$. Given an assignment π, we denote π_{label} to be the set of label-assignments $\{\lambda \mid \lambda(V) \in L_{\pi(V)}\}$.*

The new definition allows a domain to contain values having the same label but different support structures; values having both the same label and the support structure are not permitted. In some examples, we append a suffix to labels. This is not actually necessary and is used only to better differentiate between values with the same label, e.g. c_1, c_2 and c_3 in Figure 2(ii) all have the same label c. For simplicity, we use a form of cross-product representation to denote the support structure of a value. For example, value d in Figure 1(i) is represented by $(\{d\}, \{b\} \times \{f,g,h\})$. The new definition also allows multiple labels so that we can combine neighborhood interchangeable values [6] without losing any solution (unlike the standard practice where only one value is kept). For example $(\{a\}, \{x\} \times \{y\})$ and $(\{b\}, \{x\} \times \{y\})$ can be replaced by a single value $(\{a,b\}, \{x\} \times \{y\})$. Any solution involving $(\{a,b\}, \{x\} \times \{y\})$ can be converted back using label-assignment.

We define the following operations on values, analogous to the usual operations on sets. For these operations to be correct, we require constraint networks to be AC. Given the same requirement for various algorithms — for instance Maintaining Arc Consistency — and the efficiency of recent AC algorithms this property is not taxing to presume.

Definition 3 (Operations on Values). *Let a and b be two values in D_V.*

The intersection *of a and b ($a \odot b$) is a value c where $L_c = L_a \cup L_b$ and $\sigma_c(W) = \sigma_a(W) \cap \sigma_b(W)$ for all $W \in N_V$. Two values a and b are* disjoint *($a \odot b = \emptyset$) if there exists a variable $X \in N_V$ such that $\sigma_a(X) \cap \sigma_b(X) = \emptyset$. A set of values is disjoint if its members are pairwise disjoint.*

The union *of a and b ($a \oplus b$) is a value c where $L_c = L_a = L_b$ and $\sigma_c(W) = \sigma_a(W) \cup \sigma_b(W)$ for all $W \in N_V$. Union is* undefined[1] *($a \oplus b = \emptyset$) if $L_a \neq L_b$ or there exist $X, Y \in N_V$ such that $\sigma_a(X) \neq \sigma_b(X)$ and $\sigma_a(Y) \neq \sigma_b(Y)$. A* subtraction *of b from a ($a \ominus b$) is a minimal set of disjoint values $\mathscr{C}$ such that $\bigoplus(\mathscr{C} \cup \{a \odot b\}) = a$. Value a is* subsumed *by b ($a \sqsubseteq b$) if $\sigma_a(W) \subseteq \sigma_b(W)$ for all $W \in N_V$.*

For example, $(\{x,y\}, \{a,b\} \times \{c\}) \odot (\{y,z\}, \{a\} \times \{c,d\}) = (\{x,y,z\}, \{a\} \times \{c\})$, $(\{x\}, \{a\} \times \{b\} \times \{c\}) \oplus (\{x\}, \{a\} \times \{b\} \times \{d\}) = (\{x\}, \{a\} \times \{b\} \times \{c,d\})$. For subtraction, we stress that the result is a *set* of values. Consider $(\{a\}, \{d,e\} \times \{f,g\}) \ominus (\{b\}, \{d\} \times \{f\})$: two possible results[2] are $\{(\{a\}, \{e\} \times \{f,g\}), (\{a\}, \{d\} \times \{g\})\}$ and $\{(\{a\}, \{e\} \times \{f\}), (\{a\}, \{d\} \times \{e,g\})\}$. These operations preserve the local solutions of values involved.

3   Extracting Assignments

The following definitions are used to deal with the extraction process.

Definition 4 (Unit). *A value $a \in D_X$ is called a* unit *value if and only if a has exactly one support in the domain of each variable in N_X. A consistent*

---

[1] In other words, supports are only allowed to differ on one variable in the neighborhood. Union imposes this restriction so that a new value can be formed without introducing spurious solutions. Union was used in [1] but not in the algorithms here.

[2] Algorithms in this paper do not require the result of a subtraction to be unique, although it would lead to different networks. We can enforce uniqueness by imposing some ordering on the support structure.

assignment $\pi = (t_1, \ldots, t_m)$ *is called a* unit assignment *if and only if each t_i is a unit value in $\mathcal{P}|_{dom(\pi)}$.*

Given a consistent assignment π, we want to make it a unit assignment. This is so that it can be easily extracted. There are many ways to transform the domains involved so that π becomes unit. The simplest method is to solve $\mathcal{P}|_{dom(\pi)}$ for all solutions and modify domains and constraints so that a solution is represented by a single strand in the microstructure. The target unit assignment would be one of the solutions. This process is not practical as all solutions are needed and the sub-problems involved are completely replaced. We propose a dynamic method that gradually changes the network by subtracting from t_i in π the *unit value of t_i with respect to π* (defined below) until the whole assignment becomes unit.

Definition 5 (Unit Value with Respect to Assignment). *Given a consistent assignment $\pi = (t_1, \ldots, t_m)$ and a value $a = t_i$ for some $1 \leq i \leq m$, the unit value of a with respect to π (denoted by unit(a, π)) is a value u such that $L_u = L_a$ and for any X in the neighborhood of var(a)*

$$\sigma_u(X) = \begin{cases} \{t_j\} & \text{if } X = \text{var}(t_j), i \neq j \text{ and } 1 \leq j \leq m \\ \sigma_a(X) & \text{otherwise} \end{cases}$$

We explain the process using Figure 2(i)–(viii). Figure (i) depicts the initial microstructure. Suppose we want to mark the assignment $\pi = (b, c, e, f, h)$ as a nogood. This can be done by separating it from the rest of the network. The result is shown in (viii); both (i) and (viii) are equivalent in term of solution set.

The entire process could be thought of as "untangling a thread" by splitting it off one segment at a time. In (ii) for instance, the value $(\{c\}, \{a,b\} \times \{d,e\})$ in (i) is split into three values with the same label: $(\{c\}, \{b\} \times \{e\})$, $(\{c\}, \{b\} \times \{d\})$, and $(\{c\}, \{a\} \times \{d,e\})$, where $(\{c\}, \{b\} \times \{e\})$ is *unit(c, π)* for both (i) and (ii). These three values are represented as c_1, c_2, and c_3 in the figures (note that the subscript is not necessary). In this example, D_Y is chosen first and the complete order is (Y, W, Z, Y, W, U, X). Note that a domain value could be transformed more than once, and a different ordering results in a different network, although the target unit assignment is always identical.

3.1   Basic Algorithm

Algorithm 1 gives a restricted version of extraction, which requires that π is non-cyclic (defined below). We call this algorithm *extractLine()*. As we will see later, algorithm 1 also serves as a basic template for more general forms of extraction.

Definition 6 (Cyclic Assignment). *A consistent assignment π for a binary constraint network $\mathcal{P}$ is cyclic if and only if the constraint network for $\mathcal{P}|_{dom(\pi)}$ contains a cycle.*

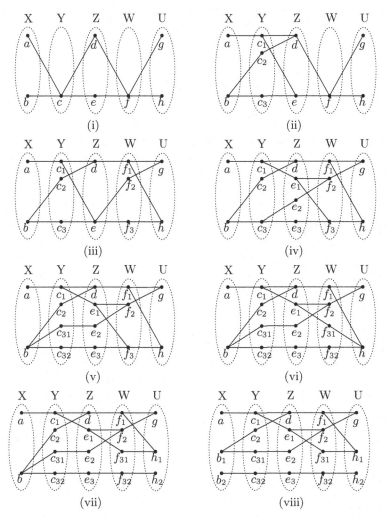

Fig. 2. Extracting a non-cyclic assignment. Figures (i)–(viii) illustrate the process of extracting (b, c, e, f, h) using the order (Y, W, Z, Y, W, U, X). Due to space limitations we do not provide details of the subtraction operation in this paper. In short, given $a \ominus b$ a subtraction algorithm will try to shed part of a by going through each variable in the neighborhood until b emerges. For instance, consider $(\{c\}, \{a,b\} \times \{d,e\}) \ominus (\{c\}, \{b\} \times \{e\})$ (from (i) to (ii)). Since $c \in D_Y$, we need to consider $N_Y = \{X, Z\}$. Suppose X is chosen first; the algorithm would divide $(\{c\}, \{a,b\} \times \{d,e\})$ into $(\{c\}, \{b\} \times \{d,e\})$ and $(\{c\}, \{a\} \times \{d,e\})$. $(\{c\}, \{b\} \times \{d,e\})$ is further divided into $(\{c\}, \{b\} \times \{d\})$ and $(\{c\}, \{b\} \times \{e\})$. Therefore, $(\{c\}, \{a,b\} \times \{d,e\}) \ominus (\{c\}, \{b\} \times \{e\}) = \{(\{c\}, \{a\} \times \{d,e\}), (\{c\}, \{b\} \times \{d\}), (\{c\}, \{b\} \times \{e\})\}$ using the order (X, Z) (this ordering has nothing to do with the ordering from line 1 of Algorithm 1). Note that $(\{c\}, \{a\} \times \{d,e\}) \oplus ((\{c\}, \{b\} \times \{d\}) \oplus (\{c\}, \{b\} \times \{e\})) = (\{c\}, \{a,b\} \times \{d,e\})$.

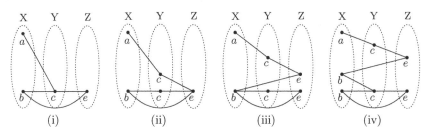

Fig. 3. Figure (i) – (iv) illustrate non-terminating transformation. Figure (i) depicts the partial microstructure induced by (b, c, e). In Figure (ii) – (iv), the processing order is (Y, Z, X). Extraction continues indefinitely for the order $(Y, Z, X, Y, Z, X, \ldots)$.

Algorithm 1. extractLine(π)

---

input: A consistent assignment $\pi = (t_1, \ldots, t_m)$
1 **while** $t_k \neq unit(t_k, \pi)$ for some $1 \leq k \leq m$ **do**
2     $R \longleftarrow t_k \ominus unit(t_k, \pi)$ /* R is a set */;
3   Replace t_k in its respective domain with $unit(t_k, \pi)$ and value(s) in R;
4   Update constraints involved with variable $var(t_k)$;

From the example in Figure 2(i)–(viii) it is not clear whether the algorithm terminates in general, since a variable domain could be repeatedly transformed. For instance, value e in (i) is unit but loses the property after its neighboring value c is split. We will prove that *extractLine()* is correct and terminates.

Definition 7 (Neighborhood Arc Consistency). *A value is* neighborhood arc consistent *(NAC) if and only if its supports are arc-consistent with one another.*

Lemma 1. *Given a constraint network and a value a, an upper-bound on the number of NAC values having the same label as a in an equivalent constraint network is d^n.*

Proof: The maximum number of solutions for the network is d^n. The network can be rearranged so that each solution is a unit assignment. Each value a in the original network participates in no more than d^n of them. Since a unit value of a unit assignment is NAC, an equivalent network can contain at most d^n NAC unit values having the same label as a.   □

Theorem 1. extractLine() *is correct and terminates on non-cycle input.*

Proof: The algorithm changes the constraint network incrementally by splitting each t_k one by one. Since operations on values, including subtraction ($\ominus$), preserve the local solutions with respect to their labels, the resulting network admits the same solutions as the original.

After the subtraction in line 2, $unit(t_k, \pi)$ takes place of t_k in the repeat loop, making t_k a unit value in the future passes. When the condition in line 1 becomes false, each component of π must be a unit value and thus π is a unit assignment according to Definition 4.

To show that the algorithm terminates, we note that there are at most $d^{|dom(\pi)|}$ NAC values having the same label as t_k according to Lemma 1. Since $t_k \neq unit(t_k, \pi)$ (in fact t_k subsumes $unit(t_k, \pi)$), each subtraction produces at least one new value r in R with the same label as t_k. The acyclic restriction on π implies there is no constraint among variables in the neighborhood of $var(t_k)$, thus r is automatically NAC. Therefore, each subtraction produces at least one NAC value. As the upper-bound on the number of NAC values having the same label as t_k and the number of components of π (which is m) are finite, the algorithm cannot keep on producing new NAC values and it must terminate within a finite number of steps.   □

Space complexity of $extractLine()$ is $O(dg)$ where g is the maximum degree of all variables in the network. This is due to the fact that the algorithm works on two values at a time (a value chosen and its unit value) and modifies only part of connecting constraint.

The algorithm may not terminate on input containing a cycle. An example of a non-terminating transformation is shown in Figure 3(i)—(iv).

3.2   Extracting Cyclic Assignments

In a non-terminating transformation that involves a cyclic assignment, values that are repeatedly split off are not part of any solution. This stems from the fact that the definition of unit value does not take into account the constraints among the neighborhood. Since $extractLine()$ operates on one segment at a time, intuitively speaking there is a chance that the segments split off will not be joined properly when the target assignment forms a cycle.

A solution to this problem is to enforce PC [7] on new values created after each subtraction. PC ensures that the segments that are split off form a connected path along the cycle. To make the algorithm terminate on cyclic input, we add the following line in the algorithm after line 4 inside the while loop: *Enforce PC on $\mathcal{P}|_{dom(\pi)}$ only on arcs involving values in R*. Figure 4 demonstrates the extraction of a cycle. We call this algorithm $extractCycle()$.

It is worth noting that having a constraint network that is already path consistent beforehand does not eliminate the need for path consistency processing in $extractCycle()$. As an example, consider $(\{d\}, \{a, b_2\} \times \{e,f\})$ in Figures 4(ii), where both (i) and (ii) are path consistent. After subtraction, d is split into three values, whose edges (b_2, d_2), (d_2, e), and (d_1, f) are path inconsistent. The reason is due to the multiplicative effect of the number of local solutions involving d ($size(d)$), while PC among its neighborhood only has the additive effect on some of those local solutions.

In order to prove that $extractCycle()$ terminates, we need to define the following notion of neighborhood path consistency.

Definition 8 (Path Consistent with Respect to Cycle). *Given a cycle involving an arc (X, Y), a tuple $(a, b) \in C_{XY}$ is path consistent with respect to*

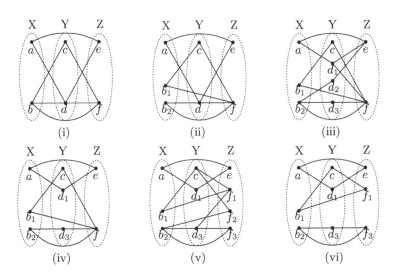

Fig. 4. Extracting (b, d, f). Figure (i) is the initial microstructure which contains three entangled solutions. In (ii) D_X is transformed; enforcing PC has no effect on the result. Figure (iii) depicts the network after D_Y is transformed; PC is later enforced, resulting in (iv). Next, D_Z is transformed as shown in (v). The final result after PC processing is shown in (vi). Notice that the three solutions are now disconnected.

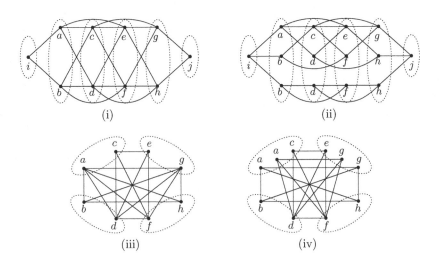

Fig. 5. More complex extraction. (i) and (iii) are the original networks while (ii) and (iv) are the results after extracting (b, d, f, h) and (a, b, g, h) respectively. Note that if (a, d, f, g) were to be extracted in (iii) instead the result in (iv) would contain three disconnected solutions.

the cycle *if and only if* (a, b) *can be extended to the whole cycle by finding values that satisfy all the constraints in the cycle.*

Definition 9 (Neighborhood Path Consistency). *A value* $a \in D_X$ *is* neighborhood path consistent (NPC) *with respect to a cycle* C *if and only if* (a, b) *is path consistent with respect to* C *for all* $b \in N_X$.

The proof of the following lemma is similar to that of of Lemma 1.

Lemma 2. *Given a value* $a \in D_X$ *and a cycle* C *involving* X, *an upper-bound on the number of NPC values with respect to* C *having the same label as* a *in an equivalent constraint network is* d^n.

In the rest of the paper when we say a value $a \in D_X$ is NPC that means it is NPC with respect to the all the cycles involving X in the input π. Non-terminating transformation involves generating non-NPC values, which can be removed by PC processing. Note that NPC implies NAC but not vice versa. We will use NPC instead of NAC in the termination proof for *extractCycle()*.

Theorem 2. extractCycle() *is correct and terminates.*

Proof: The proof is similar to that of *extractLine()* except for the algorithm termination. We will show that each subtraction will bring a certain measure closer to a finite bound. However, with no restriction on π, not all values in R of the algorithm are NPC and there is no guarantee that a subtraction will produce at least an NPC value. We will instead use the bound on the number of NPC values together with a bound on a new measure involving π.

We define the following measure $size(\pi) := \sum_{1 \leq i \leq m} \sum_{W \in N_{var(t_i)}} |\sigma_{t_i}(W)|$. Observe that after the algorithm is finished, $size(\pi)$ must be less than that of the original input assignment by a finite amount. We denote ΔL to be the value of that amount and we will use this bound along with the upper-bound in Lemma 2 (denoted by U) to prove algorithm termination. Specifically, we will show that: (1) a subtraction either produces an NPC value having the same label as t_k or decreases $size(\pi)$ by at least one. (2) a subtraction does not increase $size(\pi)$. (3) a subtraction that decreases $size(\pi)$ by one will decrease the number of NPC values by at most d^n. (In contrast, the number of NAC values in the proof of *extractLine()* does not decrease.) As a result, $size(\pi)$ decreases no more than ΔL times and the number of NPC values increases no more than $U + d^n \Delta L$ times. The algorithm terminates since these bounds are finite and either the decrease or the increase must happen after each subtraction.

We prove the three conditions as follows:

(1) Assume a subtraction does not produce an NPC value having the same label as t_k. Since t_k subsumes $unit(t_k, \pi)$, we will focus on the arc (t_k, s) where $s \neq t_i$ for any i and $s \in \sigma_{t_k}(W)$ for some W. (t_k, s) will be removed during the PC processing that follows the subtraction; otherwise it will form part of an NPC sub-value, contradicting the assumption. Thus the value of $size(\pi)$ will be decreased by the number of such arcs. (E.g (d, e) and (d, a) in Figure 4(ii)

are deleted in (iv), thereby reducing the measure by 2. We emphasize that these arcs are not path inconsistent by themselves — indeed, both arcs are PC in (ii). They are removed due to the *combination* of subtraction and PC processing.)

(2) Since subtraction preserves local solutions of t_k, if an arc is removed it would be replaced by an equivalent arc that leads to the same local solutions (e.g. (c, d) is in Figure 2(i), (c_3, d) is not in (ii), but the extra arc (b, c_2) in (ii) would keep b connected to d via c_2). An increase in the number of arcs connected to t_k and its neighborhood means an increase in the number of local solutions, which is not possible (e.g the number of arcs involving c, b, e in (i) are 4, the same number as those involving c_3, b, e in (ii) although they are different).

(3) If an arc is deleted, a number of values could lose the NPC property. We simply use the bound d^n.

Remark: we cannot rely on the reduction of $size(\pi)$ (or similar measures based on the number of links involving π) alone to prove the algorithm termination since it does not always decrease after subtraction. For instance, $size((b, c_3, e, f, h))$ in Figure 2(ii) is 12, the same as $size((b, c_3, e, f_3, h))$ in (iii).   □

Space complexity of *extractCycle()* is the same as that of *extractLine()* for the same reason. Time complexity is exponential in the worst case as a result of the upper bound used in the proof. The bounds in both proofs are admittedly very loose but they are in no way an indication of the actual number of passes. Our concern is to show that the algorithms terminate. Nevertheless, we expect the algorithm to be used in some specific context as an auxiliary routine to other algorithms, to be used occasionally, so that even if the time complexity is higher, that would not render it impractical. An algorithm which is more expensive can still be applied as a preprocessing step for some applications. We will discuss further applications in Section 5.

Another example of the process is given in Figure 6. Since the input is assumed to be consistent, we only need to enforce PC on new arcs. That is, after subtracting $unit(b, \pi)$ we need to find a support for (a, b_1), (b_1, c), (b_1, f), (e, b_2), (b_2, c), and $\{(i, j) \mid i \in \{b_1, b_2\}, j \in \{g, d\}\}$. Moreover, since PC is enforced just

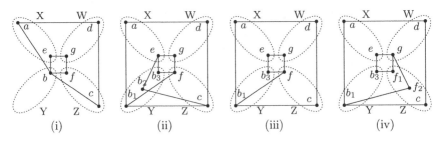

Fig. 6. Extracting (b, e, g, f). (ii) shows the result after the first subtraction. (iii) shows the result of enforcing PC — since no arc from b_2 to any value in D_W is PC, b_2 is removed ((b_2, g) has no support in Z and (b_2, d) has no support in X.) (iv) depicts the network after the next subtraction. After enforcing PC, f_2 is removed.

to eliminate non-NPC values, we do not need to record a new constraint. In this example we do not need to add C_{YW}.

Although we have shown in the proof that PC is enough for extracting cycle of any length, one might think that extracting an assignment involving k-clique would require k-consistency. This is not true since PC is used only to eliminate values that will not form proper cycles, not to solve any sub-problem. Because a k-clique can be decomposed into smaller cycles, the subtraction operation together with PC suffice in extracting any type of assignment, as long as it is consistent. Examples of more complex extraction are given in Figure 5, which includes a 4-clique and two overlapping cycles.

4   Extracting Microstructure

We can generalize the subtraction process for inputs which are microstructures rather than assignments. We formally define a microstructure as follows.

Definition 10 (Microstructure). *A microstructure of a binary constraint network $\mathcal{P}$ is a graph $\mathcal{M} = (V, E)$ in which $V \subseteq \mathcal{D}$, and if an edge (a, b) is in E then values a and b are compatible in $\mathcal{P}$. We use $\text{dom}(\mathcal{M})$ to indicate $\bigcup_{a \in V} \text{var}(a)$. A microstructure $\mathcal{M} = (V, E)$ is arc consistent if and only if given $a \in V$, $X = \text{var}(a)$, and $C_{XY} \in \mathcal{C}$, there exists $(a, b) \in E$ for some $b \in D_Y$.*

Since we define the set of vertices as a subset of domain values, the support structure of a value in a microstructure follows that of the whole network. However, we need to define a value whose support structure conforms with only edges in the microstructure.

Definition 11 (Unit Value with Respect to Microstructure). *Given an arc consistent microstructure $\mathcal{M} = (V, E)$ and a value $a \in V$, the unit value of a with respect to $\mathcal{M}$ (denoted by $\text{unit}(a, \mathcal{M})$) is a value u such that $L_u = L_a$, and for any X in the neighborhood of $\text{var}(a)$*

$$\sigma_u(X) = \begin{cases} \{b \in V | X = \text{var(b)} \text{ and } (a, b) \in E\} & \text{if } X \in \text{dom}(\mathcal{M}) \\ \sigma_a(X) & \text{otherwise} \end{cases}$$

The algorithm for extracting microstructure (*extractStructure()*) is similar to *extractCycle()* except the input is an arc consistent microstructure $\mathcal{M}$ and we use the unit value according to Definition 11 rather than Definition 5. Notice that if a microstructure $\mathcal{M}$ contains a single value per domain, then it is also an assignment. An arc consistent microstructure $\mathcal{M}$ according to Definition 10 is equivalent to the consistent assignment $\mathcal{M}$ according to Definition 1. This means an input $\mathcal{M}$ for *extractStructure()* need not contain a solution for $dom(\mathcal{M})$.

The correctness and termination proof is similar to that of *extractCycle()*. *extractStructure()* is strictly more powerful than *extractCycle()* since its input may involve more than one value from the same domain. An example of microstructure extraction is given in Figure 7.

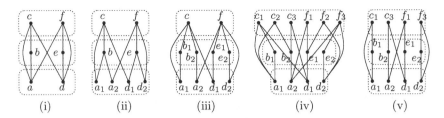

Fig. 7. Extracting cycle (a, b, c, d, e, f). (i) is the original network; (v) is the result. Notice that the cycle (a, b, c, d, e, f) is arc consistent but contains no solution (arcs (a, c) and (d, f) are not part of the cycle). From (iv) to (v) we only enforce PC on values in R of the algorithm so that arcs involving c_3 and f_1 are not checked for PC.

5 Applications

Maintaining constraint arity. A common approach taken to characterize a tuple as a nogood is to update the constraint involved so that the tuple would not be tried in the future. When constraints are table-based, this can be done by simple record-keeping. If the constraint does not exist it must be created. When the tuple is of size k, the resulting network will have at least one constraint of arity k. This change in topology is problematic for algorithms that presume the maximum arity of constraints to be bounded. Algorithms that work only for binary CSP [2] would be rendered useless when the constraint arity increases, even though the network starts out as binary. This is especially significant for distributed CSPs in which agents perform two-way communication. We resolve this problem by first extracting the target tuple and removing it by deleting one of the values involved from its domain. The extraction is done along the existing constraints and no new, higher-arity constraint is required.

Adaptive consistency [8] is one example of algorithms that produce non-binary nogoods. By using the extraction process, an initially binary constraint network will remain binary after applying adaptive consistency. Indeed, the arcs in the *ordered constraint graph* would be identical to the constraint network itself. Adaptive consistency has been used in the context of real-time constraint satisfaction [9], whose authors chose to delete domain values involved in nogoods. However, some solutions may be lost. In contrast, the removal of nogoods by extraction would allow complete solution retention without the need for extra storage space for new constraints.

As our algorithms work only for binary CSP at the moment, a possible research direction is to extend the process to directly cover non-binary constraints (though conversion to binary CSP is possible with good performance [10]). This will allow the original arity of the network to be maintained regardless of its initial value; algorithms that work only on k-ary constraints but produce nogoods involving g-ary constraint as a side-effect where $g > k$ would continue to work.

Learning nogoods by pruning. While the common approach used in nogood learning is to record each nogood as it arises [3], we can extract and discard a nogood instead. Nogoods of higher order can then be avoided without requiring a large amount of space. Although in the worst case, the resulting microstructure can be quite large, this is offset by the fact that we "learn" a nogood by pruning part of the microstructure away — in effect embedding the knowledge in the network itself — rather than by explicit memorization; the dynamic would balance the overall size as search progresses. This kind of learning is independent of variable ordering and is especially useful for non-systematic search whose completeness relies on nogood store, which becomes very large over time, such as weak-commitment search [11].

Enforcing NIC in one pass. NIC is a powerful technique and has shown to be stronger than many other types of consistency [12]. NIC requires that a value participates in a solution of its neighborhood; otherwise, the value is removed and the effect is propagated. Solving a sub-problem for a solution is a relatively expensive task however. The propagation would further increase the total cost, as every single neighborhood-inverse-inconsistent value must be removed to fully achieve NIC. In practice, NIC is rarely used and/or limited to a single pass.

We suggest a way to fully enforce NIC in one pass as follows. When a solution in the neighborhood is found, we extract it out together with the value it supports. This ensures each NIC value has only a single solution as its support after the first pass. Afterward, no sub-problem needs to be solved and further neighborhood-inverse-inconsistent values are deleted by AC propagation alone.

It remains to be seen whether it is possible to apply the idea to other higher-order consistency techniques that require multiple passes, such as Singleton Arc Consistency [13,14].

Extracting unsolvable sub-problems. In contrast to the previous use for NIC, we can extract sub-problems that are known to be unsolvable and discard them to reduce the search space. This method (using *extractStructure()*) is strictly more powerful than nogood recording since each unsolvable sub-problem may involve more than one value from the same domain. In [5], a constraint network is decomposed into disconnected sub-problems by recursively splitting *variable domains*. The result is a collection of independent constraint networks with redundant variables. Microstructure extraction is more efficient since only the target sub-problem is isolated while the rest of the network remains intact; everything is contained in just one CSP.

We can use this technique to partially enforce k-consistency by preprocessing a CSP so that any pattern in the microstructure matching known k-inconsistent sub-problems in the portfolio — for instance a pigeonhole problem — is extracted and eliminated, without the usual time complexity associated with k-consistency.

6   Conclusion

We have introduced a novel process based on value-splitting that is able to extract the target tuple/microstructure while preserving all the solutions. A number of applications are suggested. As the process involves only the most basic CSP model and is not restricted to any specialized problem or constraint, we believe that once this process is recognized more wide-ranging applications will appear. Since the extraction process increases the size of domains and constraints, it is suitable for dynamic algorithms that are able to prune parts of the microstructure away as they run, so that the increase and the decrease in size would cancel each other out.

The focus in this paper has been to show how a different view of domain values gives new approaches to existing problems. Future work is to investigate some of the implementation issues, such as ordering heuristics (i.e. how best to pick a value in line 1 of Algorithm 1), how to efficiently implement the subtraction operator, and also experimental studies.

Acknowledgment

We wish to acknowledge the support of NUS ARF grant 252-000-222-112.

References

1. Bowen, J., Likitvivatanavong, C.: Domain transmutation in constraint satisfaction problems. In: Proceedings of AAAI-04, San Jose, California (2004)
2. Bacchus, F., Chen, X., van Beek, P., Walsh, T.: Binary vs. non-binary constraints. Artificial Intelligence 140, 1–37 (2002)
3. Frost, D., Dechter, R.: Dead-end driven learning. In: Proceedings of AAAI-94, Seattle, Washington, pp. 294–300 (1994)
4. Freuder, E.C., Elfe, C.D.: Neighborhood inverse consistency preprocessing. In: Proceedings of AAAI-96, Portland, Oregon, pp. 202–208 (1996)
5. Freuder, E.C., Hubbe, P.D.: Extracting constraint satisfaction subproblems. In: Proceedings of IJCAI-95, Montreal, Canada, pp. 548–555 (1995)
6. Freuder, E.C.: Eliminating interchangeable values in constraint satisfaction problems. In: Proceedings of AAAI-91, pp. 227–233. Anaheim, California (1991)
7. Montanari, U.: Networks of constraints: fundamental properties and applications to picture processing. Information Sciences, 95–132 (1974)
8. Dechter, R., Pearl, J.: Network-based heuristics for constraint-satisfaction problems. Artificial Intelligence 34, 1–38 (1987)
9. Beck, J.C., Carchrae, T., Freuder, E.C., Ringwelski, G.: Backtrack-free search for real-time constraint satisfaction. In: Wallace, M. (ed.) CP 2004. LNCS, vol. 3258, Springer, Heidelberg (2004)
10. Samaras, N., Stergiou, K.: Binary encoding of non-binary constraint satisfaction problems: Algorithms and experimental results. Journal of Artificial Intelligence Research, 641–684 (2005)
11. Yokoo, M.: Weak-commitment search for solving constraint satisfaction problems. In: Proceedings of AAAI-94, Seattle, Washington, pp. 313–318 (1994)
12. Debruyne, R., Bessière, C.: Domain filtering consistencies. Journal of Artificial Intelligence Research, 205–230 (2001)
13. Bessiere, C., Debruyne, R.: Optimal and suboptimal singleton arc consistency algorithms. In: Proceedings of IJCAI-05, Edinburgh, U.K (2005)
14. Lecoutre, C., Cardon, S.: A greedy approach to establish singleton arc consistency. In: Proceedings of IJCAI-05, Edinburgh, U.K, pp. 199–204 (2005)
15. van Beek, P., Dechter, R.: On the minimality and global consistency of row-convex constraint networks. Journal of the ACM 42, 543–561 (1995)

Complexity of a CHR Solver for Existentially Quantified Conjunctions of Equations over Trees

Marc Meister, Khalil Djelloul*, and Thom Frühwirth

Fakultät für Ingenieurwissenschaften und Informatik
Universität Ulm, Germany

Abstract. Constraint Handling Rules (CHR) is a concurrent, committed-choice, rule-based language. One of the first CHR programs is the classic constraint solver for syntactic equality of rational trees that performs unification. We first prove its exponential complexity in time and space for non-flat equations and deduce from this proof a quadratic complexity for flat equations. We then present an extended CHR solver for solving existentially quantified conjunctions of non-flat equations in the theory of finite or infinite trees. We reach a quadratic complexity by first flattening the equations and introducing new existentially quantified variables, then using the classic solver, and finally eliminating particular equations and quantified variables.

1 Introduction

Constraint Handling Rules (CHR) [3,5,14] is a concurrent committed-choice constraint logic programming language consisting of guarded rules that transform multi-sets of constraints (atomic formulas) into simpler ones until they are solved. One of the first CHR programs is the classic constraint solver for syntactic equality of rational trees (RT) that performs unification [3,5].

Unification is concerned with making first order logic terms syntactically equivalent by substituting terms for variables. For example, the terms $h(a, f(Y))$ and $h(Y, f(a))$ can be made syntactically equivalent by substituting the constant a for the variable Y. In 1930, Herbrand [6] gave an informal description of a unification algorithm. Robinson [12] rediscovered a similar algorithm when he introduced the resolution procedure for first-order logic in 1965. Since the late 1970s, there are quasi-linear time algorithms for unification. For finite trees (Herbrand terms), see [9] and [11]. For rational trees, see [7]. These algorithms can be considered as extensions of the union-find algorithm [15] from constants to trees.

Contributions. In this paper we present two new contributions:

(1) The classic RT solver relies on a term order and its complexity (in time and space) was an open problem for a decade. We first prove its *exponential time and space complexity* for non-flat equations using any term order[1] and

---

* Funded by the DFG research project GLOB-CON.
[1] We pay the elegance of the solver, which consists of just four rules and so is more concise than most formal specifications of unification, by an exponential complexity.

F. Azevedo et al. (Eds.): CSCLP 2006, LNAI 4651, pp. 139–153, 2007.

then show its *quadratic time and space complexity* for flat equations, i.e. each equation contains at most one function symbol.

(2) We show that *existentially quantified conjunctions of non-flat equations*[2] *can be solved in Maher's theory T of finite or infinite trees [8] with a quadratic complexity* using an extension of this classic RT solver as well as a notion of *reachable variables and equations*. To the best of our knowledge, this is the first CHR solver for existentially quantified conjunctions of non-flat equations in T with *quadratic complexity*.

Organisation of the Paper. We recall the basics of Constraint Handling Rules (CHR) and present Maher's theory T of finite or infinite trees [8] in Section 2.

In Section 3 we first introduce the CHR rules of the classic RT solver which is parametrised by an order of terms and prove its termination for any term order. We then show exponential worst-case time and space complexity of RT for any term order in the case of non-flat equations. We end this section showing its quadratic complexity for flat equations.

Finally, in Section 4, we extend the classic RT solver, so that it can solve in T existentially quantified conjunctions of non-flat equations in quadratic complexity. For that, we first show that any existentially quantified conjunction of non-flat equations can be transformed into an equivalent existentially quantified conjunction of flat equations in linear time and space complexity. Then, we define the notion of reachable variables and equations and use it to remove particular quantified variables and equations. Our extension of the classic RT solver consists of only a few CHR rules.

2   Preliminaries

Readers familiar with CHR and the theory of finite or infinite trees can skip this section.

2.1   Constraint Handling Rules

Constraint Handling Rules (CHR) [3,5,14] is a concurrent, committed-choice, rule-based logic programming language. We distinguish between two different kinds of constraints: *built-in (pre-defined) constraints* which are solved by a given constraint solver and *CHR (user-defined) constraints* which are defined by the rules in a CHR program. This distinction allows one to embed and utilise existing constraint solvers as well as side-effect-free host language statements. Built-in constraint solvers are considered as black-boxes whose behaviour is trusted and that do not need to be modified or inspected.

There are two main kinds of rules:

Simplification rule Name @ $H \Leftrightarrow G \mid B$
Propagation rule   Name @ $H \Rightarrow G \mid B$

---

[2] For example, the equation $\exists X\, h(X, f(Y))$ **eq** $h(Y, f(X))$ is an existentially quantified non-flat equation with the free variable Y.

Name is an optional, unique identifier of a rule, the *head H* is a non-empty conjunction of CHR constraints, the *guard G* is a conjunction of built-in constraints, and the *body B* is a goal. A *goal (query, problem)* is a conjunction of built-in and CHR constraints. A trivial guard expression "*true* |" can be omitted from a rule. Since we do not use propagation rules in this paper, it suffices to say that they are equivalent (in the standard semantics) to a simplification rule of the form *Name* @ $H \Leftrightarrow G \mid (H \wedge B)$.

The *standard operational semantics* of CHR is given by a transition system where states are conjunctions of constraints. To the constraints in the store, rules are applied until a fix-point is reached. Note that conjunctions in CHR are considered as *multi-sets* of atomic constraints. Any rule that is applicable can be applied and rule application cannot be undone since CHR is a committed-choice language. A simplification rule $H \Leftrightarrow G \mid B$ is applicable in state $(H' \wedge C)$, if the built-in constraints C_b of C imply that H' matches the head H and imply the guard G (cf. Figure 1).

IF $H \Leftrightarrow G \mid B$ is a copy of a rule $H \Leftrightarrow G \mid B$ with new variables $\bar{X}$
AND $CT \models \forall(C_b \rightarrow \exists \bar{X}(H = H' \wedge G))$
THEN $(H' \wedge C) \rightarrowtail (B \wedge G \wedge H = H' \wedge C)$

Fig. 1. State transition for simplification rules

If applied, a simplification rule *replaces* the matched CHR constraints in the state by the body of the rule. The number of rule applications in a computation is called *derivation length*. A computation terminates in the final state when the constraint store becomes inconsistent or no rule is applicable[3].

2.2 Theory $\mathcal{T}$ of Finite or Infinite Trees

The theory $\mathcal{T}$ of finite or infinite trees, which is built on an signature containing an infinite set $\mathcal{F}$ of distinct function symbols, has as axioms the infinite set of propositions of one of the three following forms:

$$\forall \bar{X} \forall \bar{Y} \quad \neg(f(\bar{X}) \text{ eq } g(\bar{Y})) \qquad \qquad \qquad \text{[A1]}$$
$$\forall \bar{X} \forall \bar{Y} \quad f(\bar{X}) \text{ eq } f(\bar{Y}) \rightarrow \bigwedge_i X_i \text{ eq } Y_i \qquad \text{[A2]}$$
$$\forall \bar{X} \exists! \bar{Z} \quad \bigwedge_i Z_i \text{ eq } T_i[\bar{X}\bar{Z}] \qquad \qquad \qquad \text{[A3]}$$

where f and g are distinct function symbols taken from $\mathcal{F}$, $\bar{X}$ is a vector of possibly non-distinct variables X_i, $\bar{Y}$ is a vector of possibly non-distinct variables Y_i, $\bar{Z}$ is a vector of distinct variables Z_i, and $T_i[\bar{X}\bar{Z}]$ is a term which begins with an element of $\mathcal{F}$ followed by variables taken from $\bar{X}$ or $\bar{Z}$.

The forms [A1], [A2], and [A3] are also called *schemas of axioms* of the theory $\mathcal{T}$. Proposition [A1] – called *conflict of symbols* – shows that two distinct operations produce two distinct individuals. Proposition [A2] – called *explosion* – shows

---

[3] To avoid trivial non-termination, propagation is applied to the same constraints only once.

that the same operation on two distinct individuals produces two distinct individuals. Proposition [A3] – called *unique solution* – shows that for a particular form of conjunction of equations, a unique set of solutions exists in $\mathcal{T}$, e.g. the formula $\exists Z\, Z = f(Z)$ has a unique solution which is the infinite tree $f(f(f(...)))$.

Maher has axiomatised the theory $\mathcal{T}$ and shown its completeness using a decision procedure which transforms any first-order formula into a Boolean combination of quantified conjunctions of atomic formulas [8]. A more general decision procedure was recently given by Djelloul in the frame of decomposable theories [2]. Maher has also shown that the structure of finite or infinite trees and the structure of the rational trees are models of $\mathcal{T}$. A rational tree is a finite or infinite tree whose set of subtrees is finite, e.g. the infinite tree $f(f(f(...)))$ is rational as its set of subtrees $\{f(f(f(...)))\}$ is finite.

Note that $\mathcal{T}$ does not accept full elimination of quantifiers, e.g. in the formula $\exists X\, Y$ eq $f(X)$ we cannot remove or eliminate the quantifier $\exists X$. This is due to the fact that for each model M of $\mathcal{T}$ there exist instantiations $\hat{Y}$ of the free variable Y which satisfy the instantiated formula $\exists X\, \hat{Y}$ eq $f(X)$ (for example $f(1)$) and others which contradict the instantiated formula $\exists X\, \hat{Y}$ eq $f(X)$ (for example $g(1)$). As a consequence, the formula $\exists X\, Y$ eq $f(X)$ is neither true nor false in $\mathcal{T}$ and the quantifier $\exists X$ cannot be eliminated. This makes solving existentially quantified conjunctions of equations non-trivial. We show in Section 4, using the notion of *reachable variables*, how to detect whereas a quantification can be eliminated.

3   Rational Tree Equation Solver

The CHR rational tree equation solver (RT solver) is one of the first published CHR programs. However, this elegant solver depends on an order of terms.

3.1   The CHR Rational Tree Solver

The CHR program in Figure 2 solves rational tree equations [3,5]. This solver dates back to late 1993 and was revised in 1998 [14]. The underlying algorithm is similar to the one in [1], but unlike this and most other unification algorithms it uses variable elimination (substitution) only in a very limited way, if it cannot be avoided. As a consequence, the algorithm has to rely on an order on terms for termination. However, this makes termination and complexity analysis considerably harder.

We describe the RT solver where T, T_1, T_2 are meta-variables that range over arbitrary terms: *Auxiliary* built-ins allows the solver to be independent of the representation of terms. Besides *true* and *false*, we have var(T) iff T is a variable and nonvar(T) iff T is a *function term*. We rely on a total pre-order $\preceq$ on terms[4] which fulfils three properties (defined in Subsection 3.2). As usual, we

---

[4] A pre-order is reflexive and transitive, however it may not be antisymmetric, i.e. from $T_1 \preceq T_2$ and $T_2 \preceq T_1$ we cannot conclude that T_1 and T_2 are equal. A pre-order becomes an order when considering the classes of indifferent terms w.r.t. $\preceq$.

```
reflexivity    @ X eq X                <=> var(X) | true.
orientation    @ T eq X                <=> var(X), X≺T | X eq T.
decomposition  @ T1 eq T2              <=> nonvar(T1), nonvar(T2) |
                                           same_functions(T1,T2).
confrontation  @ X eq T1, X eq T2 <=> var(X), X≺T1, T1≼T2 |
                                           X eq T1, T1 eq T2.
```

Fig. 2. CHR Rational tree equation solver (RT solver)

write $T_1 \prec T_2$, iff $T_1 \preceq T_2$ and $T_2 \not\preceq T_1$. The auxiliary same_functions(T_1, T_2) leads to *false* if T_1 and T_2 have not the same function symbol and the same arity (this is called clash), otherwise a constraint lists2eq(L_1, L_2) pairwise equates the lists of arguments L_1 and L_2 of the two terms using a simple recursion:

```
lists2eq([HL1|TL1],[HL2|TL2]) <=> HL1 eq HL2, lists2eq(TL1,TL2).
lists2eq([],[])               <=> true.
```

We now explain application of each *CHR rule* of the solver:

reflexivity removes trivial equations between identical variables.

orientation reverses the arguments of an equation so that the (smaller) variable comes first.

decomposition applies to equations between two function terms. When there is a clash, same_functions leads to *false*. Otherwise, the initial equation is replaced by equations between the corresponding arguments of the terms.

confrontation replaces the variable X in the second equation X **eq** T_2 by T_1 from the first equation X **eq** T_1. It performs a limited amount of variable elimination (substitution) by only considering the l.h.s.' of equations. This rule duplicates the term T_1 and the guard makes sure that T_1 is not larger than T_2.

Due to the **confrontation** rule, the complexity of the solver is worse than linear. The intricate interaction between the **decomposition** rule and the **confrontation** rule in the case of infinite terms (cyclic terms) makes it hard to prove termination (cf. Subsection 3.2) and to determine the worst-case time complexity of the solver (cf. Subsection 3.3).

3.2 Term Order and Termination

We define a generic *term order* $\preceq$ and prove our conjecture from [10] that the RT solver terminates when used with $\preceq$.

Definition 1 (Term order). *A term order $\preceq$ is a total pre-order on terms which has the following three properties*[5]:

(i)   *For different variables X and Y, either $X \prec Y$ or $Y \prec X$.*
(ii) *Any variable is smaller than any function term.*
(iii) *Subterms are smaller than the terms that properly contain them.*

---

[5] We write $T_1 \prec T_2$, i.e. term T_1 is *smaller than* term T_2, iff $T_1 \preceq T_2$ and $T_2 \not\preceq T_1$. Recall, that a term that is not a variable is a *function term*.

This term order subsumes orders found in the literature, e.g. *term-size order* which is based on *term size*.

Definition 2 (Term size and term-size order). *The* term size $\#T$ *of a term T is the number of occurrences of variables and function symbols. A term-size order $\preceq_s$ must respect properties (i) and (ii) of Definition 1 and is based on term size by $S \preceq_s T$ iff $\#S \le \#T$ for two function terms.*

Similarly to a term-size order $\preceq_s$, we can define a *term-depth order* $\preceq_d$ which is based on the nesting depths of the terms. Note that term-size and term-depth order are not compatible, e.g. $f(f(a)) \prec_s f(a, a, a)$ in term-size order but $f(a, a, a) \prec_d f(f(a))$ in term-depth order. In previous work [10] we introduced the *term-measure order* $\preceq_m$ which is also not compatible to term-size order $\preceq_s$.

A conjunction of atomic constraints is *solved* (or in *solved normal form*) if it is either *false* or if it is of the form $\bigwedge_{i=1}^{n} X_i$ eq T_i with pairwise distinct variables $X_1, \ldots, X_n$ and arbitrary terms $T_1, \ldots, T_n$ for $n \in \mathbf{N}$. We require X_i to be different to T_j for $1 \le i \le j \le n$, i.e. if a variable occurs on the l.h.s. of an equation, it does neither occur as its r.h.s. nor as the l.h.s. or r.h.s. of any subsequent equation. By Definition 1, we can restate the conditions for the solved normal form to $X_i \prec X_{i+1}$ (for $1 \le i < n$) and $X_i \prec T_i$ (for $1 \le i \le n$).

The RT solver computes the solved form, as can be shown by contradiction: As long as a conjunction of constraints is not in solved from, at least one rule is applicable. If it is in solved form, no rule is applicable.

We now show termination of the RT solver for any term order.

Theorem 1 (Termination). *The derivation length of the RT solver, used with any term order $\preceq$, and for any given conjunction of equations is finite.*

Proof. We abstract constraints into five disjunct sorts and study the effects of rule applications.

Sort bi for the built-ins *false* or *true*,
Sort vv for equations X eq Y with two variables X and Y,
Sort vt for equations X eq T with variable X and function term T,
Sort tv for equations T eq X with function term T and variable X, and
Sort tt for equations T_1 eq T_2 with two function terms T_1 and T_2.

We give the *sort transition graph* of the RT solver in Figure 3: Each arrow visualises the effect of a rule application by removing one equation of a given sort and introducing constraints of other sorts. As the solved normal form contains only constraints of the Sorts bi, vv, or vt we indicate this with doubled-rimmed boxes. The built-in constraints of Sort bi are treated by the host language, they are never removed, and there is no arrow from Sort bi. We study the effects of each rule application in turn.

Application of reflexivity replaces the redundant equation X eq X by
       true. We visualise this by the arrow from Sort vv to Sort bi, labelled with **re**,
       in the right part of Figure 3.

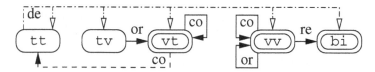

Fig. 3. Sort transition graph for the RT solver for non-flat terms

Application of `orientation` is possible to equations of two different sorts and hence two arrows are labelled with **or**. First, an equation of Sort **tv** can be changed into an equation of Sort **vt**, i.e. T `eq` $X \rightarrowtail X$ `eq` T for a variable X and a function term T. Second, when replacing Y `eq` X by X `eq` Y for two variables $X \prec Y$, both the removed and the inserted equation are of Sort **vv**. This is visualised by the self-loop labelled **or** attached to Sort **vv**.

Application of `decomposition` produces equations between the arguments of the function terms of the initial equations (or *false* for different functors). By the subterm property (iii) of $\preceq$, the arguments or the new equations are smaller than the initial arguments. We visualise this with dashed-dotted arrows, labelled **de**, from Sort **tt** to all five Sorts. The non-solid arrow-heads indicate that more than one constraint can be inserted.

Application of `confrontation` replaces one occurrence of the variable X by the value of T_1. The guard ensures that $X \prec T_1 \preceq T_2$. As long as T_1 is a variable, it gets closer from below to T_2 but can never exceed it.

– For two variables $T_1 \preceq T_2$, application of `confrontation` removes the equation X `eq` T_2 of Sort **vv** and inserts the equation T_1 `eq` T_2 of Sort **vv**. This is indicated by a self-loop of Sort **vv** labelled by **co**.
– Similarly, for a variable T_1 and a function term T_2, `confrontation` removes and inserts an equation of Sort **vt**, indicated by a self-loop of Sort **vt** labelled by **co**.

If T_1 is a function term, then so must be T_2, and we replace equation X `eq` T_2 of Sort **vt** with T_1 `eq` T_2 of Sort **tt**. As the guard requires $T_1 \preceq T_2$ we visualise this with a dashed arrow, labelled **co**.

Because Figure 3 contains all possible sort transitions, we show that all possible derivation paths, which are given by chaining the sort transitions for all equations in a given problem, are finite. First note, that the number of variables is bounded by the initial number of variables v of the problem because the solver introduces no new variables throughout the derivation.

Clearly, rule `reflexivity` applies at most *once* to a given equation. Also, rule `orientation` can apply at most once for a given equation by properties (i) and (ii) of $\preceq$. By property (i) of $\preceq$, rule `confrontation` can apply v times when term T_1 is a variable for a given equation. It remains to prove that no infinite derivation exists along the loop through Sort **tt** and Sort **vt**, respectively the number of (interleaved) rule applications of `confrontation` and `decomposition` are limited: Equations T_1 `eq` T_2 of Sort **tt**, that are created by `confrontation`, must be eventually decomposed because there is no other possible sort transition and the solved normal form contains no equations of Sort **tt**. As we have $T_1 \preceq T_2$

for equations of Sort tt which are generated by application of confrontation, subsequent application of decomposition on T_1 eq T_2 produces terms which are smaller than T_2.

Since both the number of subterms in a given problem and the number of variables v are finite, and since the term order is thus well-founded, the number of confrontation/decomposition cycles is bounded.   □

3.3   Exponential Complexity for Non-flat Equations

A closer look at Figure 3 allows us to prove that the derivation length is (at most) exponential in the problem size for any term order. By giving an exponential witness query, we also show that this result is tight for (at least) the term-size order $\preceq_s$, the term-depth order $\preceq_d$, and the measure-order $\preceq_m$ of [10]. Therefore, worst-case space and (hence) time complexity of the classic CHR constraint solver for unification is *exponential*.

Definition 3. *The* problem size $\#C$ *of a problem* $C = \bigwedge_{i=1}^{n} S_i$ eq T_i *is given by* $\sum_{i=1}^{n} \#S_i + \#T_i$.

Before proving our first main result, we present a simple, yet basic insight on the number of occurrences of *function subterms*.

Property 1. A problem containing n occurrences of function symbols contains n occurrences of function subterms.

We now study the *multi-set of function subterms* for a given problem: Application of decomposition removes two function subterms and confrontation adds function subterms when replacing a variable X by a function term T_1. Application of confrontation in the case of a variable T_1 as well as application of reflexivity or orientation are invariant to the multi-set of function subterms. As the RT solver only decomposes and copies terms, we have:

Property 2. All function subterms during a computation are function subterms of the initial problem.

Let $S_1 \preceq \cdots \preceq S_n$ be an ascending chain of all occurrences of function subterms of a problem C. The *multiplicities vector* $\langle \#S_1, \ldots, \#S_n \rangle$ is defined by the current number of occurrences of subterms during the computation, e.g. $a \preceq a \preceq f(a) \preceq g(a)$ is an ascending chain for the problem X eq $f(a) \wedge X$ eq $g(a)$ and $\langle 1, 1, 1, 1 \rangle$, $\langle 2, 0, 1, 1 \rangle$, and $\langle 0, 2, 1, 1 \rangle$ are valid initial multiplicities vectors.

Property 3. The number of confrontation/decomposition cycles, i.e. applications of confrontation with generation of an equation of Sort tt with subsequent removal by decomposition (along the dashed arrow in Figure 3), for a problem with n occurrences of function symbols is bounded by 2^n.

Replacing variable X of an equation X eq T_2 by a function term T_1 by application of confrontation adds one copy of each function subterm occurrence of term T_1 to the problem. Subsequent application of decomposition on equation T_1 eq T_2 removes one occurrence of both T_1 and T_2 from the problem.

Because the newly created equation T_1 eq T_2 of Sort tt can only be removed by decomposition, no occurrences of function subterms of T_1 eq T_2 are effected by intermediate applications of other rules and we do not need to manifest the temporarily addition of T_1 to the problem and can restrict ourselves to *all proper* function subterms of T_1. Summarising, one confrontation/decomposition cycle *adds one copy of each proper function subterm occurrence of term* T_1 and *removes one occurrence of the function subterm* T_2 from the problem, hence some entries of the multiplicities vector which are left to T_1's position increase by one and T_2's entry decreases by one.

The number of cycles is bounded by the number required to make the (canonical) initial multiplicities vector $\langle 1, \ldots, 1 \rangle$ equal to $\langle 0, \ldots, 0 \rangle$ (as then neither confrontation nor decomposition can apply). Furthermore, we use an upper bound for each cycle by increasing *all* multiplicities to the left of T_2's entry by one (and not only the effected entries of *proper subterms of* T_1) and decrease T_2's entry by one. As we increase entries to the left of a given position in the multiplicities vector, starting form the right side yields the maximal number of possible cycles: From $\langle 1, \ldots, 1 \rangle$, we arrive at $\langle 2, \ldots, 2, 0 \rangle$ after one cycle and another *two* cycles bring us (via $\langle 3, \ldots, 3, 1, 0 \rangle$) to $\langle 4, \ldots, 4, 0, 0 \rangle$. Altogether at most $\sum_{i=0}^{n-1} 2^i \leq 2^n$ many cycles suffice.

Theorem 2. *The derivation length of a problem C with problem size $\#C$ of the RT solver using any term order is bounded by $O(2^{\#C})$.*

Proof. Consider a problem C of initial size $\#C = v + n$ with v (occurrences of) variables and n (occurrences of) function symbols.

Only the transition from Sort tv to Sort tt by application of confrontation increases the problem size and the maximal number of rule applications along the dashed arrow in Figure 3 is bounded by $O(2^n) = O(2^{\#C})$, cf. Property 3. Each time a variable is replaced by a term, the problem size increases by less than $\#C$. Hence, the maximal problem size during the computation is bounded by $O(\#C\, 2^{\#C}) = O(2^{\#C})$ and the confrontation/decomposition cycle produces no more than $O(2^{\#C})$ many equations.

As application of decomposition strictly decreases the problem size, it can apply at most $O(2^{\#C})$ many times. For each equation, the self-loops can apply at most v times, altogether $O(3v\, 2^{\#C}) = O(2^{\#C})$ many rule applications, and the transition from Sort tv to Sort vt and from Sort vv to Sort bi can happen only once for each equation, altogether $O(2^{\#C})$ many rule applications.

The derivation length is hence bounded by $O(4\, 2^{\#C}) = O(2^{\#C})$.   □

We now present a witness query with exponential space complexity (for details see [10]): We apply confrontation between selected equations exhaustively before application decomposition to yield a maximal number or generated constraints. For mutually recursively defined terms

$$\mathcal{U}_i := \begin{cases} X & \text{if } i = 0 \\ f(\mathcal{L}_{i-1}, X) & \text{otherwise} \end{cases} \qquad \mathcal{L}_i := \begin{cases} X & \text{if } i = 0 \\ f(X, \mathcal{U}_{i-1}) & \text{otherwise} \end{cases}$$

the problem $C(n) = (\bigwedge_{i=1}^{n} X \ \mathtt{eq} \ \mathcal{L}_i) \wedge X \ \mathtt{eq} \ \mathcal{U}_n \wedge X \ \mathtt{eq} \ \mathcal{L}_n$ has quadratic size $\#C(n) = O(n^2)$. For any term order which satisfies that $\mathcal{L}_i$ and $\mathcal{U}_i$ are indifferent (this includes $\preceq_s$, $\preceq_d$, and $\preceq_m$) there exists a derivation which produces exponentially many equations. Precisely 2^{n+1} many equations $X \ \mathtt{eq} \ X$ are produced.

Our first main result is that the RT solver using any term order has exponential space and (hence) exponential time complexity. The derivation length is (at most) exponential for any term order (and this result is even tight, e.g. for the standard term-size order).

3.4   Quadratic Complexity for Flat Equations

We can improve the worst-case time and space complexity of the CHR rational tree solver from exponential to quadratic by simply requiring that equations are in flat form when the problem is given. For flat terms, property (ii) automatically implies (iii) of the term order.

Definition 4. *A conjunction of equations is in* flat form *if each equation contains at most one function symbol.*

For a conjunction of equations in flat form, application of $\mathtt{decomposition}$ yields equations of Sort $\mathtt{vv}$ (or *false* for different functors) as all proper subterms for flat terms are variables. Hence we can remove the arrows from Sort $\mathtt{tt}$ to Sorts $\mathtt{tt}$, $\mathtt{tv}$, and $\mathtt{vt}$ from Figure 3 and the sort transition graph for the flat problem, given in Figure 4, *lacks the intricate interaction of* $\mathtt{confrontation}$ *and* $\mathtt{decomposition}$.

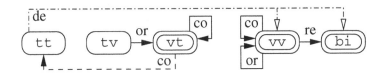

Fig. 4. Sort transition graph for the RT solver for flat terms

Theorem 3. *The derivation length of the RT solver using any term order for a* flat *problem C with problem size $\#C$ is bounded by $O(\#^2C)$.*

Proof. For each initial equations of Sort $\mathtt{vt}$, rule $\mathtt{confrontation}$ can apply at most $\#C$ times (as the number of variables is also bounded by $\#C$) and is hence bounded by $\#^2C$. The sort transition from Sort $\mathtt{vt}$ to Sort $\mathtt{tt}$ at most doubles the size of each equation and subsequent application of $\mathtt{decomposition}$ on the at most $\#C$ equations generates at most $\#C$ equations of Sort $\mathtt{vv}$. Together with initial equations of Sort $\mathtt{vv}$, rule $\mathtt{confrontation}$ applies at most $\#C$ times for each of the at most $\#C$ times many equations between two variables. $\square$

4 Solving Existentially Quantified Conjunctions of Non-flat Equations in $\mathcal{T}$

We extend the preceding RT solver for solving existentially quantified conjunctions of non-flat equations in the theory $\mathcal{T}$ of finite or infinite trees with a quadratic complexity. Solving a quantified conjunction φ of non-flat equations in $\mathcal{T}$ means to transform φ into an equivalent existentially quantified conjunction ϕ of flat equations such that ϕ is either the formula *true* or the formula *false* or a formula having at least one free variable, being neither *true* nor *false* in $\mathcal{T}$ and where the solutions of the free variables are expressed in a clear and explicit way. In particular, if φ has no free variables then ϕ is either the formula *true* or the formula *false*. The full implementation of our extended RT solver is available online at `http://www.informatik.uni-ulm.de/pm/index.php?id=139`.

4.1 Flattening Non-flat Equations in $\mathcal{T}$

The following property is easily shown in $\mathcal{T}$.

Property 4. For a conjunction of constraints $\bigwedge_{i=1}^{n} S_i$ **eq** T_i and new quantified variables $X_1, \ldots, X_n$ we have

$$\mathcal{T} \models \left(\bigwedge_{i=1}^{n} S_i \text{ eq } T_i \right) \leftrightarrow \exists X_1 \ldots \exists X_n \bigwedge_{i=1}^{n} \left(X_i \text{ eq } S_i \wedge X_i \text{ eq } T_i \right).$$

For an atomic constraint X **eq** T, with a function term $T = f(T_1, \ldots, T_n)$ and new quantified variables $X_1, \ldots, X_n$ we have

$$\mathcal{T} \models X \text{ eq } T \leftrightarrow \exists X_1 \ldots \exists X_n X \text{ eq } f(X_1, \ldots, X_n) \wedge \left(\bigwedge_{i=1}^{n} X_i \text{ eq } T_i \right).$$

This property shows that a conjunction of non-flat equations can be transformed into an existentially quantified conjunction of flat equations by adding new existentially quantified variables. In our CHR implementation, we traverse the equations of the problem once and replace each nested function symbol by a new existentially quantified variable and a new equation with that variable. For example, the formula $\exists X\, h(X, f(Y))$ **eq** $h(Y, f(X))$ is flattened to $\exists ABCX\, A$ **eq** $h(X, B) \wedge B$ **eq** $f(Y) \wedge A$ **eq** $h(Y, C) \wedge C$ **eq** $f(X)$.

In previous work [10] we showed:

Property 5. The size of the flattened problem $\#[C]$ is linear in the problem size, i.e. $\#[C] = O(\#C)$. The number of new existentially quantified variables and the number of new equations is linear in the problem size. The flattening of a problem C can be done in linear time and space w.r.t. the problem size $\#C$.

4.2 Reachable Variables and Equations

The theory $\mathcal{T}$ does not accept full elimination of quantifiers. Hence elimination of existentially quantified variables from a conjunction of equations is not evident. We present the notion of *reachable variables* and use it to detect if a quantified variable can be eliminated or not.

Definition 5. *A* basic formula *is a conjunction of equations in flat form in which all the left hand sides of the equations are variables. Let $\bar{X}$ be a vector of variables[6] and let α be a basic formula. The formula $\exists \bar{X}\,\alpha$ is called* formatted *if*

(i) α does not contain equations of the form Z eq Z or Y eq X with Z a variable, X an element of $\bar{X}$, and Y a free variable of $\exists \bar{X}\,\alpha$;

(ii) all the left hand sides of the equations of α are distinct variables.

Let us now introduce the notion of *reachable variable*:

Definition 6. *Let $\exists \bar{X}\,\alpha$ be a formatted formula. The* reachable variables *and* equations *of α from a variable X_0 (the variable X_0 can possibly belong to $\bar{X}$) are those which occur in a sub-formula of α of the form*

$$X_0 = T_0[X_1] \wedge X_1 = T_1[X_2] \wedge ... \wedge X_{n-1} = T_{n-1}[X_n] \ ,$$

where the variable X_{i+1} occurs in the term $T_i[X_{i+1}]$. The reachable variables *and equations of $\exists \bar{X}\,\alpha$ are those which are reachable in α from the free variables of $\exists \bar{X}\,\alpha$.*

Example 1. In the following formatted formula with free variable Z

$$\exists UVWXZ \text{ eq } f(U,V) \wedge V \text{ eq } g(V) \wedge W \text{ eq } f(U,V,X) \ , \tag{1}$$

the equations Z eq $f(U,V)$ and V eq $g(V)$ and the variables Z, U, and V are reachable. The equation W eq $f(U,V,X)$ and the variables W and X are not reachable. Note that the quantifications $\exists UV$ cannot be eliminated since the existence of valid instantiations of U and V in any model M of $\mathcal{T}$ depends on the instantiations of the free variable Z. In fact, if Z is instantiated by $g(0,0)$ then the preceding formula is false in M and if Z is instantiated by $f(1,g(g(g(...))))$ then the preceding formula is true in M. On the other hand, the quantification $\exists WX$ can be removed. In fact, according to axiom [A3] of $\mathcal{T}$ we have $\mathcal{T} \models \exists W\,W$ eq $f(U,V,X)$. Thus, (1) is equivalent in $\mathcal{T}$ to $\exists UVXZ$ eq $f(U,V) \wedge V$ eq $g(V)$, which is equivalent to $\exists UVZ$ eq $f(U,V) \wedge V$ eq $g(V)$.

Example 1 can help the reader to understand the following property:

Property 6. Let $\exists \bar{X}\,\alpha$ be a formatted formula. We have

$$\mathcal{T} \models (\exists \bar{X}\,\alpha) \leftrightarrow (\exists \bar{X}'\alpha')$$

---

[6] This includes the empty vector ε. Recall also that an empty conjunction of equations is always reduced to *true*.

where (i) $\bar{X}'$ is the vector of the variables of $\bar{X}$ which are reachable in $\exists \bar{X}\, \alpha$ and (ii) α' is the conjunction of the reachable equations of $\exists \bar{X}\, \alpha$.

This property simply states that non-reachable variables and equations of $\exists \bar{X}\, \alpha$ can be eliminated while the other quantified variables are linked to the instantiations of the free variables. The formatted formula $\exists \bar{X}'\, \alpha'$ is called *final solved form* of $\exists \bar{X}\, \alpha$.

4.3   Reachability in CHR

The CHR implementation of Property 6 consists of the following CHR rules.

```
r0 @ free(X), X eq T  ==> reach(X).
r1 @ reach(X), X eq T <=> nonvar(T) | reachargs(T), reach(X), X sol_eq T.
r2 @ reach(X), X eq Y <=> var(Y)    | reach(Y), reach(X), X sol_eq Y.
r3 @ reach(X), exists(X) <=> sol_exists(X), reach(X).
```

We create CHR constraints `sol_exists` and `sol_eq` for the reachable existentially quantified variables and the reachable equations of any formatted formula $\exists \bar{X}\, \alpha$. Recall that $\exists \bar{X}\, \alpha$ is a formatted formula (cf. Definition 5) for termination and correctness.

Initially, free variables of the formatted formula $\exists \bar{X}\, \alpha$ are stored in CHR constraints `free` and existentially quantified ones in `exists`. All free variables which occur as l.h.s. of an equation of the formatted formula $\exists \bar{X}\, \alpha$ are marked as reachable by rule `r0`. Rules `r1` and `r2` mark the reachable equations. For a flat term T, the built-in reachargs(T) of rule `r1` marks all arguments of T as reachable, by a simple recursion for an auxiliary constraint with rules similar to the ones for `lists2eq` of the RT solver. Rule `r3` marks the reachable quantified variables.

The complexity of this algorithm is bounded by $O(vq)$ where v is the number of distinct variables in the formatted formula $\exists \bar{X}\, \alpha$ and q is the number of equations of α. Since the left hand sides of α are distinct variables (cf. Definition 5) we have $q \leq v$ from which we deduce the following property:

Property 7. The derivation length of *reachability* is bounded by $O(v^2)$, where v is the number of distinct variables in the formatted formula.

4.4   The Solving Algorithm

To solve an existentially quantified conjunction $\exists \bar{X}\, \alpha$ of non-flat equations we apply the following algorithm.

(1) Transform $\exists \bar{X}\, \alpha$ into an equivalent existentially quantified conjunction $\exists \bar{Y}\, \beta$ of flat equations.

(2) Apply the RT solver on β using a *term-order where the variables of $\bar{Y}$ are smaller than the free variables of $\exists \bar{Y}\, \beta$.* Let δ be the obtained formula.

(3) If δ is different from *false*[7] then $\exists \bar{Y}\, \delta$ is a formatted formula whose final solved form is obtained using our CHR reachability rules.

---

[7] Note that CHR terminates immediately when *false* is inserted in the store.

Example 2. Let us solve the following formula with free variables X and V

$$\exists Y Z \, f(X) \text{ eq } f(g(X,Y)) \wedge Z \text{ eq } f(V) \wedge Z \text{ eq } f(f(Y)) \ .$$

After flattening we get

$$\exists Y Z A B C D \, A \text{ eq } f(X) \wedge B \text{ eq } f(D) \wedge A \text{ eq } B \wedge D \text{ eq } g(X,Y) \wedge$$
$$Z \text{ eq } f(V) \wedge Z \text{ eq } f(C) \wedge C \text{ eq } f(Y) \ .$$

The RT-solver returns the following formatted formula

$$\exists Y Z A B C D \, B \text{ eq } f(X) \wedge D \text{ eq } X \wedge A \text{ eq } B \wedge X \text{ eq } g(X,Y) \wedge C \text{ eq } V \wedge$$
$$Z \text{ eq } f(C) \wedge V \text{ eq } f(Y) \ .$$

We now compute the reachable variables X, V, and Y and eliminate the quantifications $\exists ABCDZ$ and the equations B eq $f(X)$, D eq X, A eq B, C eq V, Z eq $f(C)$. The final solved form of the preceding formatted formula is

$$\exists Y \, X \text{ eq } g(X,Y) \wedge V \text{ eq } f(Y) \ .$$

Note that the solutions of the free variables X and V are expressed in clear and explicit way. Moreover the quantifier $\exists Y$ could not be eliminated since Y is a reachable variable, i.e. it depends on the instantiations of X and V.

From Theorem 3, Property 7, and Property 5 we have

Theorem 4. *The derivation length of an existentially quantified non-flat problem C with problem size $\#C$ of the extended RT solver is bounded by $O(\#^2 C)$.*

5 Conclusion

The complexity of the classic CHR rational tree equation solver [3,5,14] was an open problem for more than a decade. We showed in this paper its termination and exponential complexity in time and space for any term order when handling non-flat equations, as well as its quadratic complexity for flat equations. This part of our new results extends those given in [10], which were limited to an artificial term-measure order.

Moreover we extended the solver to handle existentially quantified conjunctions of non-flat equations in quadratic time and space complexity. For that, we first flatten the equations by introducing new quantified variables, then solve the flat problem by the classic RT solver, and finally remove particular quantifiers and equations. Our new results extend those given in [10] by introducing *existentially quantified variables* and removing unnecessary quantified variables and equations using *reachability*. We are now able to express the solutions of any existentially quantified conjunction of non-flat equations in a short, clear, and explicit way in all models of the theory $\mathcal{T}$.

To the best of our knowledge, this is the first CHR solver for existentially quantified conjunctions of non-flat equations in $\mathcal{T}$ with *quadratic complexity*. Future work aims to reach a *linear complexity* solver by combining our extended RT solver with the union-find algorithm in CHR [13,4].

Acknowledgements. We thank the anonymous reviewers for their valuable comments which helped us to improve the paper.

References

1. Colmerauer, A.: Prolog and infinite trees. In: Clark, K.L., Tärnlund, S.-A. (eds.) Logic Programming, pp. 231–251. Academic Press, London (1982)
2. Djelloul, K.: Decomposable theories. J. Theory and Practice of Logic Programming (to appear)
3. Frühwirth, T.: Theory and Practice of Constraint Handling Rules. J. Logic Programming 37(1-3), 95–138 (1998)
4. Frühwirth, T.: Parallelizing union-find in Constraint Handling Rules using confluence. In: Gabbrielli, M., Gupta, G. (eds.) ICLP 2005. LNCS, vol. 3668, pp. 113–127. Springer, Heidelberg (2005)
5. Frühwirth, T., Abdennadher, S.: Essentials of Constraint Programming. Springer, Heidelberg (2003)
6. Herbrand, J.: Recherches sur la théorie de la demonstration. PhD thesis, Université de Paris, France (1930)
7. Huet, G.: Confluent reductions: Abstract properties and applications to term rewriting systems. J. ACM 27(4), 797–821 (1980)
8. Maher, M.J.: Complete axiomatizations of the algebras of finite, rational, and infinite trees. In: LICS'88, pp. 348–357, Los Alamitos (CA), USA (1988)
9. Martelli, A., Montanari, U.: An efficient unification algorithm. ACM Trans. Program. Lang. Syst. 4(2), 258–282 (1982)
10. Meister, M., Frühwirth, T.: Complexity of the CHR rational tree equation solver. In: CHR 2006, Report CW, vol. 452, pp. 77–92. K.U. Leuven, Belgium (2006)
11. Paterson, M.S., Wegman, M.N.: Linear unification. J. Computer and System Sciences 16(2), 158–167 (1978)
12. Robinson, J.A.: A machine-oriented logic based on the resolution principle. J. ACM 12(1), 23–41 (1965)
13. Schrijvers, T., Frühwirth, T.: Optimal union-find in Constraint Handling Rules. J. Theory and Practice of Logic Programming 6(1&2), 213–224 (2006)
14. Schrijvers, T., et al.: Constraint Handling Rules (CHR) web page (2007), http://www.cs.kuleuven.ac.be/~dtai/projects/CHR/
15. Tarjan, R.E., Van Leeuwen, J.: Worst-case analysis of set union algorithms. J. ACM 31(2), 245–281 (1984)

Efficient Recognition of Acyclic Clustered Constraint Satisfaction Problems*

Igor Razgon and Barry O'Sullivan

Cork Constraint Computation Centre
Department of Computer Science, University College Cork, Ireland
{i.razgon,b.osullivan}@cs.ucc.ie

Abstract. In this paper we present a novel approach to solving Constraint Satisfaction Problems whose constraint graphs are highly clustered and the graph of clusters is close to being acyclic. Such graphs are encountered in many real world application domains such as configuration, diagnosis, model-based reasoning and scheduling. We present a class of variable ordering heuristics that exploit the clustered structure of the constraint network to inform search. We show how these heuristics can be used in conjunction with nogood learning to develop efficient solvers that can exploit propagation based on either forward checking or maintaining arc-consistency algorithms. Experimental results show that maintaining arc-consistency alone is not competitive with our approach, even if nogood learning and a well known variable ordering are incorporated. It is only by using our cluster-based heuristics can large problems be solved efficiently. The poor performance of maintaining arc-consistency is somewhat surprising, but quite easy to explain.

1 Introduction

Solving real-world Constraint Satisfaction Problems (CSPs) can prove difficult for a number of applications, such as scheduling [17], frequency assignment problems [13], multi-commodity flow congestion control [10], and protein structure prediction [20]. Problems of this nature are typically posed as a network over which a set of constraints are defined. For example, for multi-commodity flow, the nodes in the network correspond to locations (warehouses and shipment locations), and the edges in the network correspond to links between locations (roads). The constraints specify capacities of links, among other things.

The complexity of these network-structured problems can be specified in terms of their graph parameters. In general, the overall problem is NP-hard; however, the complexity can be defined as being exponential in the tree-width of the network [4]. Roughly, the more tree-like the network is, the easier it is to solve. Tree-structured CSPs are important in that inference is polynomial. Inference in tree-structured CSPs has been heavily studied in the literature. Within the CSP community, it has been addressed in [7,3,5].

Because of the efficiency of inference in tree-structured CSPs, researchers have developed several methods for converting a CSP Z into a tree-structured CSP Z'. These

---

* This work was supported by Science Foundation Ireland (Grant Number 05/IN/I886).

F. Azevedo et al. (Eds.): CSCLP 2006, LNAI 4651, pp. 154–168, 2007.
© Springer-Verlag Berlin Heidelberg 2007

methods include using tree-decomposition algorithms [1], and then performing inference on Z' [11,6], or performing a structural compilation of the problem [18].

In this paper we propose a novel approach to efficiently solve clustered CSPs whose meta-constraint graph (the graph in which each cluster is replaced by a meta-variable) is *close* to a tree but not completely acyclic. This approach does not rely on a compilation of the problem. Instead we develop a class of search heuristics that can exploit the clustering in the problem to help dramatically improve the efficiency of search. Using these search heuristics we propose an efficient algorithm for solving clustered CSPs that combines constraint propagation and nogood recording. Our main result is that the proposed method *obliviously recognizes* (without spending any computational effort to do the recognition) when the meta-constraint graph of the problem becomes acyclic and, if that happens, solves the underlying CSP efficiently. Our algorithm assumes that the clustered structure of the CSP has either been identified by the user, or in a preprocessing step; usually this is a simple task to approximate manually. An algorithm that exploits our search heuristics, with nogood learning and constraint propagation, can be seen as exploiting a backdoor set of variables [19] that are not equivalent to a cycle cutset [5]. Our experiments show that MAC augmented with nogood learning and a good fail-first heuristic is significantly out-performed by a solver based on our cluster-based search heuristics when tested on clustered CSPs with a large number of variables.

The remainder of the paper is organized as follows. In Section 2 we introduce our notation. In Section 3 the basic algorithm is presented in terms of constraint propagation using forward checking, along with a deep theoretical analysis of it. We present a generalization of the algorithm in Section 4 that incorporates MAC, and also a simplified nogood recording scheme. An empirical evaluation of the algorithm is presented in Section 5. Finally, in Section 6 we make some concluding remarks and briefly outline our future work.

2 Background

2.1 Terminology

A Constraint Satisfaction Problem (CSP) Z is a triplet (V, D, C), where V is a set of *variables*, and D is the set of domains of *values* for each variable. We use the notation $\langle u, val \rangle$ to say that val is a value of variable v. A *tuple* of values is a set of values of different variables. If $\langle u, val \rangle$ belongs to a tuple T, we say that $\langle u, val \rangle$ is the *assignment* of u in T. The last item, C, of Z is the set of *constraints*. We represent each constraint $c \in C$ as a set of *forbidden* tuples of values to the variables constrained by c.[1] A *partial solution* is a tuple of values that contains *no* forbidden tuple as a subset. A *complete solution* or just a *solution* is a partial solution assigning all the variables of V. To *solve* a CSP is to find one of its solutions or to prove that no solution exists.

Let V' be a set of variables assigned by a forbidden tuple T. We say the variables of V' are *constrained*. If all the forbidden tuples have length 2, we get a *binary* CSP. The forbidden tuples of a CSP are often referred to as *conflicts*.

---

[1] We employ an unusual definition of a constraint because it is more convenient for our discussion.

A tuple T of values is a *nogood* if it is a forbidden tuple or if any extension of it to a tuple that assigns all the variables contains a forbidden tuple. In particular, if a partial solution is a nogood then it cannot be extended to a full solution.

2.2   The FC-EBJ Algorithm

In this paper we design CSP *solvers* that, besides solving a CSP, achieve some theoretical guarantee on their runtime. We assume the reader is familiar with constraint solvers like Forward Checking (FC) [9] and Maintaining Arc-Consistency (MAC) [16]. All the constraint solvers maintain additional data structures where they keep the results of intermediate computation. We introduce some additional terminology related to these data structures.

The *run* of a constraint solver consists of a number of iterations where it enumerates partial solutions in order to extend them to a full solution or to prove that none exists. If we consider a particular iteration occurring during the execution of a solver then the partial solution P that the solver tries to extend at that iteration is called the *current partial solution*. At the considered iteration, some values may be temporarily removed from their domains because the solver detected that they cannot be utilized to extend the current partial solution. The values that are *not* removed from their domains are called *valid* values at that iteration. The set of valid values of variable v is called the *current domain* of v.

As stated above, a value $\langle u, val \rangle$ is removed from the current domain of u because $P \cup \{\langle u, val \rangle\}$ cannot be extended to a full solution, in other words, it contains a nogood T. The set $T \setminus \{\langle u, val \rangle\}$ (which is a subset of the current partial solution) is called an *eliminating explanation* of $\langle u, val \rangle$.

The only reason why a complete constraint solver removes a value from its current domain is that it has found (implicitly or explicitly) an eliminating explanation for it. Some algorithms, like FC and MAC, do not record eliminating explanations explicitly but there are other algorithms, like DBT [8] or MAC-DBT [12], that do. A simple way to transform FC into an algorithm maintaining eliminating explanations is to introduce the following three modifications (in all the items below P denotes the current partial solution).

- If a value $\langle u, val \rangle$ is removed because $P \cup \{\langle u, val \rangle\}$ contains a forbidden tuple T, $\langle u, val \rangle$ is associated with the eliminating explanation $T \setminus \{\langle u, val \rangle\}$.
- If a value $\langle u, val \rangle$ is removed by backtracking caused by the empty current domain of some variable v, the eliminating explanation of $\langle u, val \rangle$ is the union of elimination explanations of all values of a variable v minus $\langle u, val \rangle$.
- A removed value is restored in the current domain only when its eliminating explanation becomes *obsolete*, i.e. it is no longer a subset of P.

It follows from the last property that this algorithm possesses the ability to backjump [14]. For example, if the domain of the currently unassigned variable v is wiped out, but the last assignment in the eliminating explanations of the values of v is $\langle u, val \rangle$ assigned 10-th to the last, the domain of v will remain empty until 10 consecutive backtracks (or one backjump of length 10) are performed. Thus the algorithm may be viewed as a modification of FC-CBJ [14], which maintains separate "conflict sets" for every value. We term this algorithm FC-EBJ, replacing Conflict-Directed Backjumping by Explanation-Directed Backjumping. FC-EBJ is almost analogous to the CCFC- algorithm [2] with two differences. First, the description of CCFC- is not based on eliminating explanations which makes it less convenient for our discussion. Second, CCFC- is formulated for binary constraints only, while FC-EBJ can solve any CSP.

3 Recognizing Clustered Acyclic CSPs

Consider a CSP whose variables are partitioned into clusters specified by the user. The given CSP is *clustered acyclic* if, after contraction of the clusters so that they become single vertices, the constraint graph of the given CSP is transformed into a tree. A clustered acyclic CSP can be solved with a complexity equivalent to that of solving the CSP associated with the largest cluster multiplied by the number of clusters, which is much faster than traditional backtracking if the clusters are small.

There are CSPs corresponding to real-world problems that can be divided into relatively small clusters such that the constraint graph resulting from the contraction of these clusters is close to a tree, some of which were discussed in Section 1. Therefore, it would be worthwhile to design a search algorithm that can *recognize* that the CSP induced by the *current* domains of unassigned variables is a clustered acyclic CSP and, if that happens, solves the given CSP efficiently. A straightforward approach to do that is to apply a checking procedure after every instantiation made by the search algorithm. However, this approach requires additional time for checking the desired property and thus might be not worthwhile. In this section we propose an alternative approach for recognizing clustered acyclic CSPs. In particular, we show that FC-EBJ combined with nogood learning and guided by a specially designed variable-ordering heuristic *obliviously recognizes* clustered acyclic CSPs without spending additional time or memory.

The starting point for the design of the proposed method was the observation that FC-EBJ solves a tree-like CSP polynomially if the variables are assigned in a depth-first search (DFS) manner with respect to the constraint graph of the given CSP. Next, we observed that FC-EBJ can benefit from exploring the constraint graph in a DFS-like manner even if the given CSP is not acyclic. In particular, let P be the current partial solution maintained by FC-EBJ. If the residual CSP (the result of projecting the underlying CSP to the unassigned variables and removing values incompatible with P) is acyclic then FC-EBJ takes polynomial time to check whether P can be extended to a full solution or it is a nogood. What is most interesting is that FC-EBJ neither "knows" that the residual CSP is acyclic nor applies any additional effort to recognize that. The algorithm just proceeds exploring the search space as before and the polynomial time complexity of exploring the residual CSP "comes" automatically. We say that FC-EBJ with a special ordering heuristic *obliviously recognizes* acyclic CSPs. Our last observation was that if this ordering heuristic is combined with nogood learning then the resulting version of FC-EBJ obliviously recognizes acyclic clustered CSPs.

It is worth noting that the proposed method differs from the traditional approach of compiling clusters and treating them as meta-variables with the set of solutions to the clusters as meta-values[2]. Instead, we implicitly apply a "lazy" compilation. That is, solutions to clusters may be learned as nogoods if the underlying CSP is hard and, in the worst case, the proposed algorithm has the same time and space complexity as the traditional method. However, if the underlying CSP is not too hard, it might be solved by the proposed algorithm without any compilation at all. This "compile when needed" paradigm results in huge time savings, as witnessed by the work of Bayardo and Miranker [3].

We begin the description of the proposed method from the formal definition of an acyclic clustered CSP.

Definition 1 (Clustered Acyclic CSP). *Let $Z = (V, D, C)$ be a CSP and let $SV = \{V_1, \dots V_l\}$ be a partition of V. Assume that each constraint in C constrains variables that belong to at most two different elements of SV. Let H be a constraint graph of Z i.e. a hypergraph with the vertices corresponding to V and the hyperedges corresponding to the subsets of V, which are scopes of the constraints of C. Consider the graph H' obtained from H by contracting the sets $V_1, \dots V_l$ to single vertices $v_1, \dots v_l$, removing the loops, and replacing the multiple edges by single edges. (According to our assumption all the edges of the resulting graph are binary). We denote by H' the clustered graph of Z with respect to SV. We say that Z is a clustered acyclic CSP with respect to SV, if H' is acyclic.*

Figure 1 illustrates the notion of acyclic clustered CSP. The dots represent CSP variables. All the constraints are binary. The constrained variables are connected by lines. The set of clusters $SV = \{\{V_1, V_2, V_3\}, \{V_4, V_5, V_6\}, \{V_7, V_8, V_9\}\}$. Clearly, as a result of contraction of these clusters, we get an acyclic constraint graph.

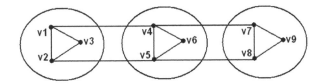

Fig. 1. Illustration of an acyclic clustered CSP

To proceed, we need to extend our notation regarding acyclic clustered CSPs. Let Z and SV be as in Definition 1. We denote by $M(Z, SV)$ the *maximal number of partial solutions* of a CSP created by the variables of V_i (i.e. the CSP consisting of the variables of V_i, their domains and the forbidden tuples assigning the variables of V_i only) among all $V_i \in SV$. Let v be a variable of Z. We denote by $Cl(v)$ the cluster of SV that v belongs to.

Now, we combine the FC-EBJ solver (Section 2.2) with a nogood learning procedure. Everytime a value $\langle u, val \rangle$ is removed, and the removal is justified by an eliminating explanation P', the resulting algorithm records the nogood $P' \cup \{\langle u, val \rangle\}$ in the store of nogoods. Nogoods are *discarded* from the store according to the following algorithm.

---

[2] In this setting recognizing clustered acyclic CSPs is equivalent to recognizing acyclic CSPs.

Let $(\langle v_1, val_1 \rangle, \ldots, \langle v_k, val_k \rangle)$ be a nogood recorded in the store. Assume the assignments are listed in the chronological order of their appearance in the current partial solution and the assignments that do not belong to the current partial solution are listed at the end ordered arbitrarily. Assume that for some $l < m < k$, $Cl(v_{l+1}) = \ldots = Cl(v_m)$ and $Cl(v_{m+1}) = \ldots = Cl(v_k)$ but $Cl(v_m) \neq Cl(v_k)$. Then the nogood is stored until the assignment $\langle v_l, val_l \rangle$ becomes obsolete (removed from the current partial solution). In other words, a nogood is stored until its obsolete part assigns variables of 3 or more clusters. All the time the nogood is stored, it is considered by the algorithm as a forbidden tuple. We call the resulting solver FC-EBJ-NL (NL is the abbreviation of Nogood Learning).

Proving the properties of FC-EBJ-NL, we extensively use the notions defined below (recall that P always denotes the current partial solution).

Definition 2 (Actual Forbidden Tuple). *Let S be a forbidden tuple (either original or dynamically acquired). We say that S is actual if it satisfies one of the following two conditions:*

- *There is a removed value $\langle u, val \rangle \in S$ such that $S \setminus \{\langle u, val \rangle\}$ is the eliminating explanation for $\langle u, val \rangle$.*
- *All the variables of $V(S \setminus P)$ (recall that P is the current partial solution) are unassigned and all the values of $S \setminus P$ belong to the current domains of their variables.*

Thus a forbidden tuple is *actual* if it is either used for pruning domain values or still may be violated by an extension of the current partial solution.

Definition 3 (Current CSP and Current Set of Clusters). *Let P be the current partial solution of FC-EBJ-NL applied to a CSP Z with a set of clusters SV.*

- *The current CSP Z' has the set of variables $V(Z) \setminus V(P)$. The domain of each variable $u \in V(Z')$ is the current domain of this variable. The constraints are defined as follows: for each actual forbidden tuple Q of Z including the values of Z', the tuple $Q \setminus P$ is forbidden in Z'. (The definition includes the forbidden tuples dynamically stored during the solution process at the considered moment of execution of FC-EBJ-NL.)*
- *The current set of clusters SV' is obtained from SV by removing the variables assigned by P from each cluster of SV.*

The proposed class of ordering heuristics includes any ordering heuristic that satisfies the following three conditions.

1. The first variable is selected arbitrarily.
2. Let u be the last assigned variable and let V_i be the cluster containing u. Assume that V_i contains an unassigned variable. Then the variable assigned next to u belongs to V_i.
3. Assume that each cluster is either fully assigned or fully unassigned. Let V' be an unassigned cluster *sharing* an actual forbidden tuple S (either original or dynamically acquired) with the *chronologically latest* possible assigned cluster V'' (in the

sense that S involves variables of both clusters). Then select any variable of V'. If there is no cluster V' as above, i.e. no assigned cluster shares an active forbidden tuple with an unassigned cluster, then select an arbitrary variable.

The last item of the above description has a much simpler formulation for the binary case: select a variable whose domain values are removed because of conflicts with the *latest possible* assigned cluster. We call the class of heuristics satisfying the above conditions LCC, which is the acronym for Last Conflicting Cluster.

The main theorem regarding FC-EBJ-NL using a LCC heuristic is as follows.

Theorem 1 (Main Result). *Assume that FC-EBJ-NL guided by a LCC heuristic is applied to a CSP Z with a collection of clusters SV. Let P be the current partial solution, let Z' be the current CSP, and let SV' be the current set of clusters. Assume that Z' is a clustered acyclic CSP with respect to SV'. Then the algorithm checks whether P can be extended to a full solution (and returns such a solution if yes, or refutes P if no) in time $M(Z', SV')^2 * (|SV'| - 1)$.*

Lemma 1. *Assume that FC-EBJ-NL is applied to a CSP Z with a set of clusters SV and consider an intermediate iteration that occurs during the processing of Z. Let P be the current partial solution, Z' be the current CSP, and SV' be the current set of clusters. Then the number of iterations spent by FC-EBJ-NL in order to check whether P can be extended to a full solution equals the number of iterations spent by FC-EBJ-NL in order to solve Z'.*

Proof of Lemma 1 is elementary but quite lengthy, we omit the proof due to space constraints. According to Lemma 1, Theorem 1 is implied by the following theorem.

Theorem 2. *Let Z be a clustered acyclic CSP with respect to a set of clusters SV. Then FC-EBJ-NL guided by a LCC heuristic solves Z performing at most $M(Z, SV)^2 * (|SV| - 1)$ backtracks.*

Proof of Theorem 2. To prove Theorem 2, we recall the way the clusters of Z are assigned by the LCC heuristic. After the full assignment of the first cluster V_1, the heuristic selects the next cluster V_2 such that V_2 "shares" nogoods with the assignments of V_1, the next cluster V_3 is selected on the same principle with respect to V_2 and, if impossible, with respect to V_1. Proceeding analogously, having assigned clusters $V_1, \ldots, V_k$, the algorithm selects the next cluster V_{k+1} so that it is constrained with the assignments of an already assigned cluster V_i, $1 \le i \le k$, such that i is as large as possible.

However, the above process is not always successful. It may happen that after assigning clusters $V_1, \ldots, V_k$ no unassigned cluster is constrained with the assignments of these clusters. In this case, the clusters $V_1 \ldots V_k$ constitute a *link* and the algorithm starts a new link. The partial solution P consists of a number of consecutive links. Each link has the *root*, i.e. the cluster, which is fully assigned first. Each cluster V_i of a link which is not the root of the link has a *parent*, i.e. the cluster which *certified* the addition of V_i to the link (in the last item of the description of the LCC heuristics, cluster V'' certifies the addition of cluster V' to a link). This structure of the current partial solution allows us to prove the following lemma.

Lemma 2. *Assume that FC-EBJ-NL guided by the LCC heuristic discards a value* $\langle u, val \rangle$ *at some state during its execution and associates it with the eliminating explanation* T. *Then* $T' = T \cup \{\langle u, val \rangle\}$ *assigns at most two clusters. Moreover, if a cluster* V' *other then* $Cl(u)$ *is assigned by* T' *then*

- $Cl(u)$ *is not assigned by the current partial solution or* V' *is the parent of* $Cl(u)$;
- V_i *and* $Cl(u)$ *are adjacent in the cluster graph of* Z.

Proof of Lemma 2 (sketch). We consider the chronological sequence of eliminating explanations generated by FC-EBJ-NL and apply induction on this sequence. Let T' be the first eliminating explanation generated by FC-EBJ-NL. Clearly, T' is a forbidden tuple of Z, hence the adjacency condition holds automatically for this case. According to the structure of Z', T' assigns at most two clusters. Assume that $Cl(u)$ is partially assigned and let us show that if T' assigns *exactly* two clusters then it assigns, besides $Cl(u)$, the parent V_i of $Cl(u)$. Assume that T' assigns a cluster V_j other than V_i. Observe first that V_i and V_j are both fully assigned.

If V_i and V_j belong to the same link then, according to the behaviour of the LCC heuristic, there is a path from V_i to V_j in the clustered tree which includes only assigned clusters of their link. Since T' is the first nogood generated by FC-EBJ-NL, the next cluster is selected as sharing an *original* forbidden tuple with the previous clusters. Therefore clusters that get to the same link induce a connected subgraph of the clustered graph of Z. On the other hand, there is a path from V_i to V_j through $Cl(u)$, which is not fully assigned (u is unassigned at the moment $\langle u, val \rangle$ is removed). It follows that there are two paths from V_i to V_j in the clustered tree, a contradiction.

If V_i and V_j belong to different links then T' is a forbidden tuple connecting clusters V_j and $Cl(u)$. If at the moment the link of V_j is fully assigned, all the values of T' are valid then we get contradiction to that fact that the new link is started. If some of the values of T' are removed, they are associated with eliminating explanations involving the variables of the link of V_j or an earlier link. In any case, it is a contradiction that $Cl(u)$ belongs to a link assigned *later* than the link of V_j.

Consider now a nogood T' with the assumption that the eliminating explanations generated prior to it satisfy the conditions of the lemma. The following claims hold at the moment of generating T' (their proofs are omitted due to space constraints).

1. Each cluster of an existing link is adjacent to its parent in the clustered graph of Z.
2. Let V_j and V_k be two clusters assigned by a stored actual nogood S and assigned by the current partial solution. Then one of them is the parent of the other.

Given the above observations, we prove the present lemma for each possible way a new nogood T' may be generated.

- T' **is an original or acquired forbidden tuple.** By the structure of Z' and the induction assumption, T' assigns at most two clusters. If T' assigns *exactly* two clusters, assume that $Cl(u)$ is assigned by the current partial solution and we prove that the other assigned cluster is the parent V_i of $Cl(u)$. If we assume that the other assigned cluster is some V_j that belongs to another link then we get a contradiction, analogously to the basic case. If we assume that the other assigned cluster belongs to the same link as $Cl(u)$ then applying the first observation from the above list and the induction assumption, we derive the existence of a cycle in the clustered graph of Z.

- T' **is the union of eliminating explanations of a variable** v **with the empty domain.** If $Cl(v)$ is assigned by the current partial solution then, by the second observation from the above list, the nogood associated with each value of v assigns the parent of V (if it assigns any cluster besides $Cl(v)$). Clearly, the union of the eliminating explanations has the same form. If $Cl(v)$ is not assigned by the current partial solution then assume that there are two nogoods T_1 and T_2 associated with values of v which assign distinct clusters, both different from $Cl(v)$. By the induction assumption, both these clusters are adjacent to $Cl(v)$ in the clustered graph of Z hence they belong to different links. Assume that T_1 involves an *earlier assigned* link. This means that there is an actual forbidden tuple involving variables of that link in contradiction to the fact that a new link has been started.   □

Thus Lemma 2 proves that nogoods discovered by FC-EBJ-NL assign at most two clusters and if two clusters are assigned, they are adjacent in the clustered tree of the underlying CSP. For each pair of clusters there are at most $M(Z, SV)^2$ possible nogoods assigning those clusters. Also there are $|SV| - 1$ pairs of adjacent clusters. Hence, the number of nogoods of the above type is at most $M(Z, SV)^2 * (|SV| - 1)$. Each nogood, once recorded in the nogood store, is never removed from there because, to be removed, a nogood must assign at least three different clusters. It follows that no nogood is discovered by a backtrack more than once. Consequently, FC-EBJ-NL spends at most $M(Z, SV)^2 * (|SV| - 1)$ backtracks solving a clustered acyclic CSP. Thus we have proved Theorem 2 and, as a result, Theorem 1.   ■

4   Enhancements of the Basic Recognition Algorithm

In this section we discuss two modifications of the FC-EBJ-NL algorithm presented in Section 3. The first modification introduces achieving arc-consistency as the constraint propagation method. The second modification simplifies the procedure of nogood learning.

4.1   Recognizing Acyclic CSP Together with MAC

Let us formulate the MAC-EBJ-NL algorithm. In addition to the propagation performed by FC-EBJ-NL, MAC-EBJ-NL removes *unsupported* values. More precisely, let u and v be two unassigned variables. If MAC-EBJ-NL detects that a value $\langle u, val \rangle$ of the current domain of u is inconsistent with all the values of the current domain of v (in the sense that $\langle u, val \rangle$ has binary conflicts with all these values either existing or produced by propagation of non-binary forbidden tuples), the following operations are performed: computing S, the union of eliminating explanations of all the values of v; removing $\langle u, val \rangle$ and associating it with eliminating explanation S; recording $S \cup \{\langle u, val \rangle\}$ in the nogood store.

We consider two versions of MAC-EBJ-NL. In the first version, which we call Clustered MAC-EBJ-NL, variables u and v, as above, must belong to the same cluster, i.e.

$Cl(u) = Cl(v)$. In other words, $\langle u, val \rangle$ is not removed if it is unsupported by variable v that does not lie in the same cluster as u does. The second version of MAC-EBJ-NL performs *full* MAC, checking any pair (u, v) of constrained variables.

It turns out that Clustered MAC-EBJ-NL preserves the property of FC-EBJ-NL, that is, recognizes a clustered acyclic CSP Z performing at most $M(Z, SV)^2 * (|SV| - 1)$ backtracks. Observe that the nogood T resulting from removing $\langle u, val \rangle$ unsupported by a variable v consists of $\langle u, val \rangle$ and the union of eliminating explanations of all the removed values of v. Applying the same induction principle as in Lemma 2, we may assume that the union of the eliminating explanations assigns, besides $Cl(v)$, at most one cluster V'. If $Cl(u) = Cl(v)$ then clearly the resulting nogood T preserves the same property.

Surprisingly enough, full MAC-EBJ-NL does not preserve this property of FC-EBJ-NL. The reason is that when a value $\langle u, val \rangle$ is removed as being unsupported by a variable v from *another* cluster, the nogood produced as a result does not necessarily satisfy the conditions stated in Lemma 2. In particular, if $Cl(v)$ is an assigned cluster and $Cl(u)$ is not the parent of $Cl(v)$ then, as a result of removing $\langle u, val \rangle$, the obtained nogood might contain three clusters: $Cl(v)$, the parent of $Cl(v)$, and $Cl(u)$. One can show that because of the above phenomenon, nogoods assigning an arbitrary number of clusters may be generated and a bad ordering of values can cause an exponential number of generated nogoods. We omit the complete proof of this fact due to space constraints.

4.2 Simplifying the Nogood Learning Procedure

FC-EBJ-NL uses quite a complicated mechanism for removing obsolete nogoods. This mechanism is useful for the proof of Theorem 1 but it might be difficult for the practical implementation. In this section we present an alternative nogood learning procedure and show that FC-EBJ combined with this procedure efficiently recognizes acyclic clustered CSPs.

The proposed mechanism of learning allocates K "slots" for nogoods given some predefined constant K. Initially all these slots are empty. The first learned nogood is stored in the first slot, the next nogood is stored in the second one, then in the third and so on. No nogood is discarded until all the slots are occupied. Once this happens, the first nogood stored after that goes to the first slot erasing the nogood stored there before that, the next one goes to the second slot with the same effect and so on. In other words, this is a mechanism of nogood learning that uses a store of a constant size and removes the "oldest" nogood in case of overflow.

The following proposition, whose proof is omitted due to space constraints, makes the above storage useful for our purposes.

Proposition 1. *Fix some two states during the execution of a constraint solver which uses the above nogood recording mechanism. Assume that at most K nogoods are stored by the solver between these two states. Then none of these K nogoods is erased until the second state is reached.*

Given a CSP Z with the set of clusters SV, set $K = M(Z, SV)^2 * (|SV| - 1)$. Assume that at some moment during the execution of FC-EBJ-NL, the current CSP is a clustered acyclic one. By Theorem 1, in order to recognize an acyclic CSP, FC-EBJ-NL detects at most K nogoods provided that no one of them is discarded during the process of recognition. According to the Proposition 1, the presented mechanism of nogood learning guarantees they are not discarded. It follows that acyclic clustered CSPs can be efficiently recognized given the presented nogood learning mechanism which is much simpler than the one described in Section 3.

5   Experimental Evaluation

In this section we report results of an empirical evaluation carried out in order to assess the practical merits of our approach to efficiently recognize acyclic clustered CSPs. We performed experiments on *clustered random problems*. We generated these problems using the following parameters: number of variables (num_var); domain size (dom_size); size of cluster ($size\_cluster$), which always divides num_var; the number of clusters ($num\_clusters = num\_var/size\_cluster$); additional connectivity ($add\_connect$), a number from 0 to 99; cluster density (cp_1); cluster tightness (cp_2); external density (ep_1); external tightness (ep_2).

Given these parameters, a CSP is generated by the following process:

1. **Create the graph of clusters** GC. Create $num\_cluster$ vertices $v_1, \ldots,$ $v_{num\_clusters}$. Then generate a tree on these vertices as follows. First the tree consists of vertex v_1. Assume that the current tree consists of vertices $v_1, \ldots, v_i$. The new vertex v_{i+1} is connected to one of the existing vertices selected uniformly at random. Having created the spanning tree, additional edges are introduced as follows. For each pair of non-adjacent vertices of the tree, a number between 0 to 99 is chosen at random. If this number is smaller than $add\_connect$ then the corresponding edge is introduced.
2. **Fill each cluster with variables.** For each cluster, $size\_cluster$ variables are generated. One of these variables per cluster has $dom\_size/2$ values in its domain (or $(dom\_size-1)/2$ in case dom_size is odd). The domain size of the rest of the variables in a cluster is dom_size. The reason of introducing one variable per cluster with a smaller domain size will be clear when we present the experimental results.
3. **Create conflicts within clusters.** This is done analogously to the well-known generator of Prosser [15] given parameters cp_1 and cp_2 which are analogous to the parameters p_1 and p_2 introduced by Prosser, respectively. In particular, the parameter cp_1 serves as the probability that there is a constraint between two particular variables in the given cluster. If the constraint exists, the parameter cp_2 determines the probability of a conflict between two values of the given variables.
4. **Create conflicts between variables of different clusters.** For each pair of variables u and v that belong to different clusters *connected by an edge in* GC, parameter ep_1 serves as the probability that there is a constraint between u and v. If the constraint exists, the parameter ep_2 serves as the probability that the given pair of values, one from u, the other from v, are conflicting.

In our experiments we took $num\_var = 100$, $dom\_size = 10$, $size\_cluster = 10$. That is, the generated CSPs have 100 variables partitioned into 10 clusters, 10 variables in each one. The additional connectivity is selected to be 5, that is, the topology of the graph of clusters is close to a tree. We performed tests for two values of cluster density: 80% and 90%. We selected just these values because clusters must be dense "by definition". For cluster density of 80% we chose the cluster tightness of 30%, for cluster density of 90% the chosen cluster tightness is 25%. The external density is always 10%. The external tightness is the varied parameter.

The rest of the section is divided into 2 subsections. In the first subsection we compare MAC-based algorithms, in the second subsection FC-based algorithms are compared.

5.1 Comparison of MAC-Based Algorithms

We have tested the following MAC-based algorithms:

- MAC-EBJ, i.e. the solver that maintains arc-consistency, records eliminating explanations for the removed values but employs no nogood learning. The smallest-domain first (a Fail-First – FF) heuristic [9], i.e. the heuristic that selected the smallest domain first, was used to guide the search performed by MAC-EBJ.
- MAC-EBJ-NL with the FF heuristic (referred as MAC-EBJ-NL-FF), i.e. the solver like the previous one with the only difference that a nogood learning mechanism is employed.
- MAC-EBJ-NL with a LCC heuristic (referred as Full MAC-EBJ-NL-LCC). The ties are broken by the FF heuristic, i.e. every time when a set of variables may be selected by the LCC heuristic, the one with the smallest domain is selected from this set.
- The same solver as the previous one but performing Clustered MAC (referred as Clustered MAC-EBJ-NL-LCC).

The algorithms above are tested on the two sets of instances presented in the introductory part of the section. We now explain why the instances are designed so that there is one variable per cluster with a smaller domain. This is done in order to "fool" the FF heuristic. Without that trick, the FF heuristic guides the search pretty much like an LCC heuristic which make the comparison of LCC and FF heuristics senseless.

Figures 2 and 3 show the behaviour of the last three solvers on sets of instances presented at the introductory part of the section. All the solvers use the simplified nogood learning mechanism described in Section 4.2. The size of the store is 10000 nogoods. The computational effort is measured in the number of backtracks. For each set of parameters, the result is obtained as the average of 50 runs. A run is stopped if it takes more than 50000 backtracks.

According to our experiments, MAC-EBJ (the solver mentioned first) was unable to solve most of the problems in the allocated number of iterations, hence we do not illustrate its behaviour in the figures.

One can see that both Full and Clustered MAC-EBJ-NL-LCC essentially outperform MAC-EBJ-NL-FF. In particular, for density of 90% at the phase transition region, Full MAC-EBJ-NL-LCC performs about 8 times better and Clustered MAC-EBJ-NL-LCC performs about 3 times better then MAC-EBJ-NL-FF. Note also that Clustered MAC-EBJ-NL-LCC outperforms MAC-EBJ-NL-FF doing much less constraint propagation.

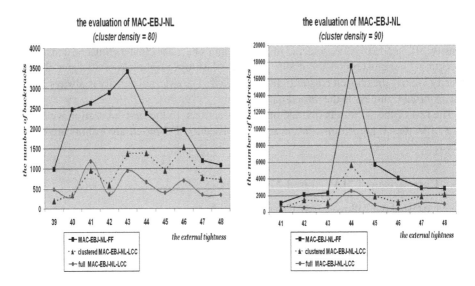

Fig. 2. Comparison of MAC-based algorithms, 80% density

Fig. 3. Comparison of MAC-based algorithms, 90% density

Specifically, the Clustered MAC-EBJ-NL-LCC only propagates within its current cluster, but this is sufficient to improve upon MAC-EBJ-NL-FF. However, when we use our cluster-based search ordering heuristics we can fully propagate using MAC and gain some additional improvements, up to a factor of 3, over Clustered MAC-EBJ-NL-LCC.

5.2 Comparison of FC-Based Algorithms

We have tested the following FC-based algorithms:

- FC-EBJ guided by FF heuristic that does not do any nogood learning. This solver is referred to as FC-EBJ-FF-WL (the last two letters abbreviate 'Without Learning').
- FC-EBJ-NL guided by the FF heuristic (referred as FC-EBJ-NL-FF).
- FC-EBJ-NL guided by the LCC heuristic (referred as FC-EBJ-NL-LCC). The breaking of ties is the same as in the case of MAC.

The algorithms that used the nogood learning mechanism are presented in Section 4.2. The algorithms have been tested on almost the same sets of instances as MAC-based algorithms with the only difference that for the cluster density 90% we have reduced the $add\_connect$ parameter to 4 because for value 5 of that parameter, all the algorithms took too long time to solve the resulting instances. The computational effort, as before, is measured in the number of backtracks, for each set of parameters 50 runs were applied with taking the average, the algorithms were stopped if they did more than 10^8 consistency checks. The experimental results are presented in Figure 4 and 5.

One can clearly observe that the nogood learning considerably improves the performance of the algorithms: both FC-EBJ-NL-FF and FC-EBJ-NL-LCC do much better than FC-EBJ-FF-WL. In particular, FC-EBJ-NL-LCC is about 12 times better than

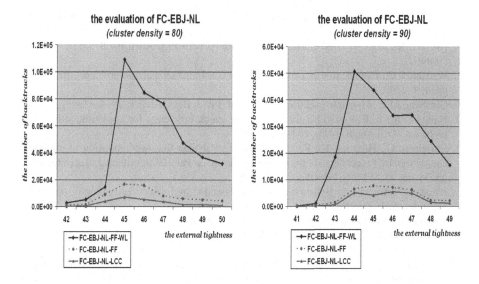

Fig. 4. Comparison of FC-based algorithms, 80% density

Fig. 5. Comparison of FC-based algorithms, 90% density

FC-EBJ-FF-WL for both considered groups of instances at the phase transition. These results correlate with the results of Bayardo and Miranker [3] on restricted nogood learning. Also FC-EBJ-NL-LCC is better than FC-EBJ-NL-FF but the rate of improvement is smaller than in the case when FC is replaced by MAC. It follows that the class of LCC heuristics are best applicable for MAC-based solvers.

6   Conclusion

It is well known that constraint graphs that are high clustered can be very challenging to solve using search-based methods. On the other hand, methods based on compilation require exponential space in the worst case. In this paper we have presented a novel algorithm for solving clustered CSPs efficiently. This algorithm can detect and exploit a tractable case automatically. Experimental results show the efficiency of our approach on large clustered CSPs.

References

1. Arnborg, S.: Efficient algorithms for combinatorial problems on graphs with bounded, decomposability–a survey. BIT 25(1), 2–23 (1985)
2. Bacchus, F.: Extending forward checking. Principles and Practice of Constraint Programming 35–51 (2000)
3. Bayardo, R.J., Miranker, D.P.: An optimal backtrack algorithm for tree-structured constraint satisfaction problems. Artif. Intell. 71(1), 159–181 (1994)
4. Bodlaender, H.L.: A tourist guide through treewidth. Acta Cybernetica 11, 1–21 (1993)

5. Dechter, R., Pearl, J.: Network-based heuristics for constraint-satisfaction problems. Artif. Intell. 34(1), 1–38 (1987)
6. Dechter, R., Pearl, J.: Tree clustering for constraint networks. Artif. Intell. 38(3), 353–366 (1989)
7. Freuder, E.C.: Complexity of k-tree structured constraint satisfaction problems. In: AAAI, pp. 4–9 (1990)
8. Ginsberg, M.L.: Dynamic backtracking. J. Artif. Intell. Res (JAIR) 1, 25–46 (1993)
9. Haralick, R.M., Elliott, G.L.: Increasing tree search efficiency for constraint satisfaction problems. Artif. Intell. 14(3), 263–313 (1980)
10. Harrelson, C., Hildrum, K., Rao, S.: A polynomial-time tree decomposition to minimize congestion. In: Proceedings of the ACM symposium on Parallel algorithms and architectures, pp. 34–43. ACM Press, New York (2003)
11. Jegou, P., Terrioux, C.: Hybrid backtracking bounded by tree-decomposition of constraint networks. Artif. Intell. 146(1), 43–75 (2003)
12. Jussien, N., Debruyne, R., Boizumault, P.: Maintaining arc-consistency within dynamic backtracking. In: Dechter, R. (ed.) CP 2000. LNCS, vol. 1894, pp. 249–261. Springer, Heidelberg (2000)
13. Koster, A.M.C.A., van Hoesel, S.P.M., Kolen, A.W.J.: Solving frequency assignment problems via tree-decomposition. Research Memoranda 036, METEOR (1999)
14. Prosser, P.: Hybrid Algorithms for the Constraint Satisfaction Problem. Computational Intelligence 9, 268–299 (1993)
15. Prosser, P.: Binary constraint satisfaction problems: Some are harder than others. In: ECAI, pp. 95–99 (1994)
16. Sabin, D., Freuder, E.C.: Contradicting conventional wisdom in constraint satisfaction. In: ECAI, pp. 125–129 (1994)
17. Sadeh, N.M., Sycara, K.P., Xiong, Y.: Backtracking techniques for the job shop scheduling constraint satisfaction problem. Artif. Intell. 76(1-2), 455–480 (1995)
18. Weigel, R., Faltings, B.: Compiling constraint satisfaction problems. Artificial Intelligence 115, 257–289 (1999)
19. Williams, R., Gomes, C.P., Selman, B.: Backdoors to typical case complexity. In: IJCAI, pp. 1173–1178 (2003)
20. Xu, J., Jiao, F., Berger, B.: A tree-decomposition approach to protein structure prediction. In: 2005 IEEE Computational Systems Bioinformatics Conference (CSB'05), pp. 247–256. IEEE Computer Society Press, Los Alamitos (2005)

Cost-Based Filtering for
Stochastic Inventory Control

S. Armagan Tarim[1], Brahim Hnich[2], Roberto Rossi[3], and Steven Prestwich[3]

[1] Department of Management, Hacettepe University, Turkey
armagan.tarim@hacettepe.edu.tr
[2] Faculty of Computer Science, Izmir University of Economics, Turkey
brahim.hnich@ieu.edu.tr
[3] Cork Constraint Computation Centre, University College, Cork, Ireland
{r.rossi,s.prestwich}@4c.ucc.ie

Abstract. An interesting class of production/inventory control problems considers a single product and a single stocking location, given a stochastic demand with a known non-stationary probability distribution. Under a widely-used control policy for this type of inventory system, the objective is to find the optimal number of replenishments, their timings and their respective order-up-to-levels that meet customer demands to a required service level. We extend a known CP approach for this problem using a cost-based filtering method. Our algorithm can solve to optimality instances of realistic size much more efficiently than previous approaches, often with no search effort at all.

1 Introduction

Inventory theory provides methods for managing and controlling inventories under different constraints and environments. An interesting class of production/inventory control problems is the one that considers the single-location, single-product case under non-stationary stochastic demand. Such a problem has been widely studied because of its key role in Material Requirement Planning [30].

We consider the following inputs: a planning horizon of N periods and a demand d_t for each period $t \in \{1, \dots, N\}$, which is a random variable with probability density function $g_t(d_t)$. In the following sections we will assume without loss of generality that these variables are normally distributed. We assume that the demand occurs instantaneously at the beginning of each time period. The demand we consider is non-stationary, that is it can vary from period to period, and we also assume that demands in different periods are independent. A fixed delivery cost a is considered for each order and also a linear holding cost h is considered for each unit of product carried in stock from one period to the next. We assume that it is not possible to sell back excess items to the vendor at the end of a period. As a service level constraint we require the probability to be at least a given value α that at the end of every period the net inventory will

F. Azevedo et al. (Eds.): CSCLP 2006, LNAI 4651, pp. 169–183, 2007.

not be negative. Our aim is to find a replenishment plan that minimizes the expected total cost, which is composed of ordering costs and holding costs, over the N-period planning horizon, satisfying the service level constraints.

Different inventory control policies can be adopted to cope with the described problem. A policy states the rules to decide when orders have to be placed and how to compute the replenishment lot-size for each order. For a discussion of inventory control policies see [29]. One of the possible policies that can be adopted is the replenishment cycle policy, (R, S). Under the non-stationary demand assumption this policy takes the form (R^n, S^n) where R^n denotes the length of the nth replenishment cycle and S^n the order-up-to-level for replenishment (Fig. 1). In this policy a wait-and-see strategy is adopted, under which the actual order quantity Q_n for replenishment cycle n is determined only after the demand in former periods has been realized. The order quantity Q_n is computed as the amount of stock required to raise the closing inventory level of replenishment cycle $n - 1$ up to level S^n. In order to provide a solution for our problem under the (R^n, S^n) policy we must populate both the sets R^n and S^n for $n = \{1, \ldots, N\}$.

Early works in this area adopted heuristic strategies such as those proposed by Silver [20], Askin [2] and Bookbinder & Tan [5]. The first complete solution method for this problem was introduced by Tarim & Kingsman [23], who proposed a certainty-equivalent Mixed Integer Programming (MIP) formulation for computing (R^n, S^n) policy parameters. Empirical results showed that such a model is unable to solve large instances, but Tarim & Smith [24] introduced a more compact and efficient Constraint Programming (CP) formulation of the same problem that showed a significant computational improvement over the MIP formulation.

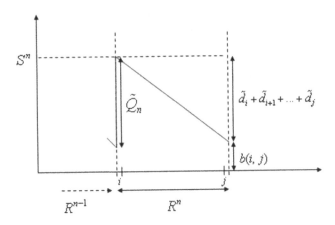

Fig. 1. (R^n, S^n) policy. R^n denotes the set of periods covered by the nth replenishment cycle; S^n is the order-up-to-level for this cycle; $\tilde{Q}_n$ is the expected order quantity; $\tilde{d}_i + \tilde{d}_{i+1} + \ldots + \tilde{d}_j$ is the expected demand; $b(i, j)$ is the buffer stock required to meet service level α

This paper extends Tarim & Smith's work, retaining their model but augmenting it with a *cost-based filtering* method to enhance domain pruning. Cost-based filtering is an elegant way of combining techniques from CP and Operations Research (OR) [7,8]: OR-based optimization techniques are used to remove values from variable domains that cannot lead to better solutions. This type of domain filtering can be combined with the usual CP-based filtering methods and branching heuristics, yielding powerful hybrid search algorithms. Cost-based filtering is a novel technique that has been the subject of significant recent research, but to the best of our knowledge has not yet been applied to stochastic inventory control. In the following sections we will show that it can bring a significant improvement when combined with the state-of-the-art CP model for stochastic inventory control.

The paper is organized as follows. Section 2 describes the CP model introduced by Tarim & Smith. Section 3 describes a relaxation that can be efficiently solved by means of a shortest path algorithm, and produces tight lower bounds for the original problem which is used to perform further cost-based filtering. Section 4 evaluates our methods. Section 5 draws conclusions and discusses future extensions.

2 A CP Model

In this section we review the CP formulation proposed by Tarim & Smith [24]. First we provide some formal background related to constraint programming. Recall that a Constraint Satisfaction Problem (CSP) [1,6] is a triple $\langle V, C, D \rangle$, where V is a set of decision variables each with a discrete domain of values $D(V_k)$, and C is a set of constraints stating allowed combinations of values for subsets of variables in V. Finding a solution to a CSP means assigning values to variables from the domains without violating any constraint in C. We may also be interested in finding a feasible solution that minimizes (maximizes) the value of a given objective function over a subset of the variables. Constraint solvers typically explore partial assignments enforcing a local consistency property using either specialized or general purpose propagation algorithms. Such propagation algorithms in general exploit some structure of the problem to prune decision variable domains in more efficient ways.

The stochastic programming (for a detailed discussion on stochastic programming see [33]) formulation for the (R^n, S^n) policy proposed in [5] is

$$\min \ E\{TC\} = \int_{d_1} \int_{d_2} \cdots \int_{d_N} \sum_{t=1}^{N} (a\delta_t + h \cdot \max(I_t, 0)) \, g_1(d_1) g_2(d_2) \ldots g_N(d_N)$$

$$\mathrm{d}(d_1)\mathrm{d}(d_2)\ldots\mathrm{d}(d_N) \tag{1}$$

subject to, for $t = 1 \ldots N$

$$I_t + d_t - I_{t-1} \geq 0 \tag{2}$$

$$I_t + d_t - I_{t-1} > 0 \Rightarrow \delta_t = 1 \tag{3}$$

$$Pr\{I_t \geq 0\} \geq \alpha \tag{4}$$

$$I_t \in \mathbf{Z}, \quad \delta_t \in \{0,1\} \tag{5}$$

Each decision variable I_t represents the inventory level at the end of period t. The binary decision variables δ_t state whether a replenishment is fixed for period t ($\delta_t = 1$) or not ($\delta_t = 0$). The objective function (1) minimizes the expected total cost over the given planning horizon.

The respective CP formulation proposed in [24] is

$$\min \; E\{TC\} = \sum_{t=1}^{N} \left(a\delta_t + h\tilde{I}_t \right) \tag{6}$$

subject to, for $t = 1 \ldots N$

$$\tilde{I}_t + \tilde{d}_t - \tilde{I}_{t-1} \geq 0 \tag{7}$$

$$\tilde{I}_t + \tilde{d}_t - \tilde{I}_{t-1} > 0 \Rightarrow \delta_t = 1 \tag{8}$$

$$Y_t \geq j \cdot \delta_j \quad j = 1, \ldots, t \tag{9}$$

$$element\,(Y_t, b(\cdot, t), H_t) \tag{10}$$

$$\tilde{I}_t \geq H_t \tag{11}$$

$$\tilde{I}_t, H_t \in \mathbf{Z}^+ \cup \{0\}, \quad \delta_t \in \{0,1\}, \quad Y_t \in \{1, \ldots, N\} \tag{12}$$

where $b(i,j)$ is defined by

$$b(i,j) = G^{-1}_{d_i + d_{i+1} + \ldots + d_j}(\alpha) - \sum_{k=i}^{j} \tilde{d}_k$$

The $element(X, list[], Y)$ constraint [31] enforces a relation such that variable Y represents the value of element at position X in the given list. $G_{d_i + d_{i+1} + \ldots + d_j}$ is the cumulative probability distribution function of $d_i + d_{i+1} + \ldots + d_j$. It is assumed that G is strictly increasing, hence G^{-1} is uniquely defined.

Each decision variable $\tilde{I}_t$ represents the expected inventory level at the end of period t. Each $\tilde{d}_t$ represents the expected value of the demand in a given period t according to its probability density function $g_t(d_t)$. The binary decision variables δ_t state whether a replenishment is fixed for period t ($\delta_t = 1$) or not ($\delta_t = 0$). The objective function (6) minimizes the expected total cost over the given planning horizon. The two terms that contribute to the expected total cost are ordering costs and inventory holding costs. Constraint (7) enforces a no-buy-back condition, which means that received goods cannot be returned to the supplier. As a consequence of this the expected net inventory at period t must be no less than the expected net inventory in period $t+1$ plus the expected demand in period t. Constraint (8) expresses the replenishment condition. We have a replenishment if the expected net inventory at period t is greater than the expected net inventory in period $t + 1$ plus the expected demand in period t. This means that we received some extra goods as a consequence of an order. Constraints (9,10,11) enforce the required service level α. This is done by specifying the minimum buffer stock required for each period t in order to ensure that, at the end of each and every time period, the probability that the net inventory will not be negative is at least α. These buffer stocks, which are stored in matrix $b(\cdot, \cdot)$, are pre-computed following the approach suggested in [23]. In this approach the authors transformed a chance-constrained model, that is a model where constraints on some random variables have to be maintained at prescribed levels of probability, in a completely deterministic one. For further details about chance-constrained programming see [32]. More specifically the authors developed a certainty-equivalent constraint for each chance constraint that enforces the required service level at the end of each replenishment cycle.

2.1 Computational Complexity

The chance-constrained problem presented in [5] for the (R^n, S^n) policy under stochastic demand is PSPACE-complete as shown in [25]. We assume that negative orders are not allowed, so that if the actual stock exceeds the order-up-to-level for that period, this excess stock is carried forward and not returned to the supply source. However, such occurrences are regarded as rare events and accordingly the cost of carrying the excess stock and its effect on the service level of subsequent periods is ignored. Under these assumptions the chance-constrained problem can be expressed by means of the certainty-equivalent model we presented, where buffer stocks for each possible replenishment cycle are computed independently. In [4] Florian et. al. gave an overview for the complexity of deterministic production planning. In particular they established NP-hardness for this problem under production cost (composed of a fixed cost and a variable unit cost), zero-holding cost and arbitrary production capacity constraint. They also extended this result by considering other possible cost functions and capacity constraints. Polynomial algorithms are discussed in the same paper for specific cases. Among these they cited Wagner and Whitin's [27] work, where the infinite capacity deterministic production planning problem is solved in polynomial time. Wagner and Whitin's algorithm relies upon their *Planning Horizon Theorem*, which exploits the fact that the feasible region is a closed bounded convex set and that the cost function is concave [4], thus the minimum value for such an objective function is achieved at one of the extreme points of this set. The special structure of the set allows a simple characterization of the production plans corresponding to its extreme points. The core insight proposed by Wagner and Whitin is the fact that in the search for the optimal policy it is sufficient to consider programs in which at period t one does not both place an order and bring in inventory. Their *Planning Horizon Theorem* states that if it is optimal to incur a setup cost in period t, when periods $1, \ldots, t$ are considered in isolation, then we may retain this decision for the N period model without losing optimality. Therefore it is possible to adopt an optimal program for period $1, \ldots, t - 1$ considered separately. It is easy to see that the certainty-equivalent model described

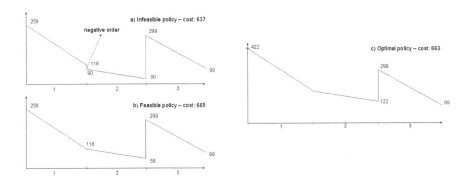

Fig. 2. a) The infeasible policy and its cost obtained by means of Wagner and Whitin algorithm. b) The respective feasible policy and its cost. c) The optimal policy and its cost obtained by using our certainty-equivalent model.

in the former section is an over-constrained version of the infinite capacity deterministic production planning problem. The additional constraints in the model we presented enforce buffer stocks for each replenishment cycle. Since we have buffer stocks, the last period of a replenishment cycle usually requires a positive inventory in our certainty-equivalent model, so it is possible that an order is placed even if the inventory level is not null. Therefore the simple characterization of optimal programs proposed by Wagner and Whitin cannot be applied, since buffer stock carried from former periods may affect the cost of subsequent programs.

We shall now show, by using a counter-example, that Wagner and Whitin's algorithm cannot be applied to the over-constrained problem, for which therefore no polynomial algorithm is known. Let us consider a 3-period planning horizon. The demand is normally distributed in each period with coefficient of variation 0.3. The mean values of the demand are respectively 240, 60, 200 for periods 1, 2, 3. The required service level is 95%, the ordering cost is 130 and the holding cost is 1. The required buffer stock levels for the possible replenishment cycles are $b(1,1) = 118, b(2,2) = 30, b(3,3) = 99, b(1,2) = 122, b(2,3) = 103, b(1,3) = 157$. The optimal policy can be easily obtained by solving our certainty-equivalent model (Fig. 2 - c). Such a policy fixes orders in periods 1 and 3 and its cost is 663. Following the same reasoning in [27] (Table 1) the optimal plan for period 1 is to order (entailing an ordering cost of 130 and a holding cost of 118). Two possibilities must be evaluated for period 2; order in period 2, and use the best policy for period 1 considered alone (at a cost of $160 + 248 = 408$); or order in period 1 for both periods, and carry inventory into period 2 (at a cost of $130 + 122 \cdot 2 + 60 = 434$). The better policy appears to be the first one, but it is actually the second one. In period 3, if the algorithm in [27] worked there would be three alternatives: order in period 3, and use the best policy for period 1 and 2 considered alone (at a cost of $229 + 408 = 637$); or order in period 2 for the latter two periods and use the best policy for period 1 considered alone (at a cost of $536 + 248 = 784$); or order in period 1 for the entire three periods (at a cost of

Table 1. Wagner and Whitin algorithm steps. In the optimal policy row only the last period is shown; $\underline{3}$ indicates that the optimal policy for periods 1 through 3 is to order in period 3 to satisfy $\tilde{d}_3$ and adopt an optimal policy for periods 1 through 2 considered separately.

Period	1	2	3
	248	408	637
		434	784
			1061
Minimum cost	248	408	637
Optimal policy	$\underline{1}$	2	$\underline{3}$

1061). The policy obtained by Wagner and Whitin's algorithm is therefore the best among these three, which places an order in period 1, 2 and in period 3 at a cost of 637. Unfortunately this policy is infeasible because it requires a negative order quantity in period 2 (Fig. 2 - a). The respective feasible policy that places orders in the same periods has a higher cost of $130 \cdot 3 + 118 + 58 + 99 = 665$ (Fig. 2 - b). This counter-example shows that Wagner and Whitin's algorithm is not suitable for our deterministic equivalent problem.

2.2 Domain Pre-processing

In [24] the authors showed that a CP formulation for computing optimal (R^n, S^n) policies provides a more natural way of modeling the problem. In contrast to the equivalent MIP formulation the CP model requires fewer constraints and provides a nicer formulation. However, the CP model has two major drawbacks. Firstly, in order to improve the search process and quickly prove optimality, tight bounds on the objective function are needed. Secondly, even when it is possible to compute *a priori* the maximum values that such variables can be assigned to, these values (and therefore the domain sizes of the $\tilde{I}_t$ variables) are large. The domain size value is equal to the amount of stock required to satisfy subsequent demands until the end of the planning horizon, meeting the required service level when only a single replenishment is scheduled at the beginning of the planning horizon.

To address the domain size issue, Tarim & Smith proposed two pre-processing methods in order to reduce the size of the domains before starting the search process, by exploiting properties of the given model and of the (R^n, S^n) policy. Method I computes a cost-based upper bound for the length of each possible replenishment cycle $T(i,j)$, starting in period i, for all $i,j \in \{1, \ldots, N\}$, $i \leq j$. Note that $T(i,j)$ denotes the time span between two consecutive replenishment periods i and $j+1$. Method I therefore identifies sub-optimal replenishment cycle lengths allowing a proactive off-line pruning, which eliminates all the expected inventory levels that refer to longer sub-optimal replenishment cycles. Method II employs a dynamic programming approach, by considering each period in an iterative fashion and by taking into account in each step two possible course of action: an order with an expected size greater than zero is placed or no order (equivalently an order with a null expected size) is placed in the considered period within our planning horizon. The effects of these possible actions in each step are reflected in the decision variable domains by removing values that are not produced by any course of action.

3   Cost-Based Filtering by Relaxation

The CP model as described so far suffers from a lack of tight bounds on the objective function. We now propose a relaxation for our model to compute a valid lower bound at each node of the search tree. We first show that the CP model can be reduced to a Shortest Path Problem if we relax constraints (7,8) for replenishment periods. That is for each possible pair of replenishment cycles $\langle T(i, k-1), T(k, j) \rangle$ where $i, j, k \in \{1, \ldots, N\}$ and $i < k \leq j$, we do not consider the relationship between the opening inventory level of $T(k, j)$ and the closing inventory level of $T(i, k-1)$.

This corresponds to allowing negative replenishments, or the ability to sell stock back to the supplier. In this way we obtain a set $\mathcal{S}$ of $N(N+1)/2$ possible different replenishment cycles. Our new problem is to find an optimal set $\mathcal{S}^* \subset \mathcal{S}$ of consecutive disjoint replenishment cycles that covers our planning horizon at the minimum cost. We will show that the optimal solution to this relaxation is given by the shortest path in a graph from a given initial node to a final node where each arc has a specific cost.

If N is the number of periods in the planning horizon of the original problem, we introduce $N + 1$ nodes. Since we assume, without loss of generality, that an order is always placed at period 1, we take node 1, which represents the beginning of the planning horizon, as the initial one. Node $N + 1$ represents the end of the planning horizon. Recall that $b(i, j)$ denotes the minimum buffer stock level required to satisfy a given service level constraint during the replenishment cycle $T(i, j)$. For each possible replenishment cycle $T(i, j - 1)$ such that $i, j \in \{1, \ldots, N + 1\}$ and $i < j$, we introduce an arc (i, j) with associated cost $Q(i, j)$, where

$$Q(i, j) = a + h \sum_{k=i+1}^{j-1} (k - i)d_k + h(j - i)b(i, j - 1) \qquad (13)$$

The cost of a replenishment cycle is the sum of two components: a fixed ordering cost a that is charged at the beginning of the cycle when an order is placed, and a variable holding cost h charged at the end of each time period within the replenishment cycle and proportional to the amount of stocks held in inventory. Since we are dealing with a one-way temporal feasibility problem [27], when $i \geq j$, we introduce no arc. The connection matrix for such a graph, of size $N \times (N + 1)$, can be built as shown in Table 2.

By construction the cost of the shortest path from node 1 to node $N + 1$ in the given graph is a valid lower bound for the original problem, as it is a solution of the relaxed problem. Furthermore it is easy to map the optimal solution for the relaxed problem, that is the set of arcs participating to the shortest path, to a solution for the original problem by noting that each arc (i, j) represents a

Table 2. Shortest Path Problem Connection matrix

	1	2	...	j	...	$N+1$
1	–	$Q(1,2)$	...	$Q(1,j)$	...	$Q(1,N+1)$
⋮	–	–	⋱	⋮	⋱	⋮
i	–	–	–	$Q(i,j)$	...	$Q(i,N+1)$
⋮	–	–	–	–	⋱	⋮
N	–	–	–	–	–	$Q(N,N+1)$

replenishment cycle $T(i, j-1)$. The feasibility of such a solution with respect to the original problem can be checked by verifying that it satisfies every relaxed constraint. To find a shortest path, and hence a valid lower bound, we use an improved Dijkstra algorithm that finds a shortest path in $O(n^2)$ time, where n is the number of nodes in the graph. Details on efficient implementations of the Dijkstra algorithm can be found in [19]. Usually Dijkstra's algorithm does not apply any specific rule for labeling when ties are encountered in sub-path lengths. This is incorrect if we pre-process decision variable domains as described in [24]. In fact pre-processing Method I in [24] relies upon an upper bound for optimal replenishment cycle length. When a replenishment period $i, i \in \{1, ..., N\}$ is considered, it looks for the lowest j s.t. $j \geq i$ after which it is no longer optimal to schedule the next replenishment. This means that, if other policies exist that share the same expected cost, only the one that has shorter, and obviously more, replenishment cycles will be preserved by Method I. Therefore, when the algorithm is implemented in this filtering approach, we need to introduce a specific rule for node selection in order to make sure that, when more optimal policies exist, our modified algorithm will always find the one that has the highest possible number of replenishment cycles (i.e. the shortest path with the highest possible number of arcs). As there is a complete ordering among nodes, we can easily implement this rule when labeling by always choosing as ancestor the node that minimizes the distance from the source and that has the highest index.

We now see how to use this relaxation during the search process when a partial solution is provided. If in a given partial solution a decision variable δ_k, $k \in \{1, ..., N\}$ has been already set to 0, then we can remove from the network every inbound arc to node k and every outbound arc from node k. This prevents node k from being part of the shortest path, and hence prevents period k from being a replenishment period. On the other hand, if $\delta_k = 1$ then we split the planning horizon into two at period k, thus obtaining two new subproblems $\{i, ..., k-1\}$ and $\{k, ..., j\}$. We can then separately solve these two subproblems by relaxing them and applying Dijkstra's algorithm. Note that the action of splitting the time span is itself a relaxation; in fact it means overriding constraints (7,8) for $t = k$. It follows that the cost of the overall solution obtained by merging the two subproblem solutions is again a valid lower bound for the original problem. Let $R(i, j)$ denote the required minimum opening inventory level in period i, $i \in \{1, \ldots, N\}$, to meet demand until period $j + 1$, where $R(i,j) = b(i,j) + \sum_{t=i}^{j} \tilde{d}_t$. We can characterize when such a bound is an exact one: when the solutions of the two subproblems are both feasible with respect to the original model and the condition

$$b(i, k - 1) \leq R(k, j) \tag{14}$$

is satisfied, the solution obtained by merging those for the independent subproblems is both feasible and optimal for the original problem. We have shown how to act when each of the possible cases, $\delta_i = 1$ and $\delta_i = 0$ is encountered. It is now possible at any point of the search in the decision tree to apply this relaxation to compute valid lower bounds. It is also possible to extend this cost-based filtering by considering not only the δ_t variable assignments, but also the $\tilde{I}_t$ variable assignments. In fact, when we compute the cost of a given replenishment cycle $T(i, j-1)$ (arc (i, j) in the matrix), we can also consider the current assignments for the closing inventory levels $\tilde{I}_t$ in the periods of this cycle. Since all the closing inventory levels of the periods within a replenishment cycle are linearly dependent, given an assignment for a decision variable $\tilde{I}_t$ we can easily compute all the other closing inventory levels in the cycle using $\tilde{I}_t - \tilde{d}_t - \tilde{I}_{t-1} = 0$, which is the inventory conservation constraint when no order is placed in period t. When the closing inventory levels in a replenishment cycle $T(i, j - 1)$ are known it is easy to compute the overall cost associated with this cycle, which is by definition the sum of the ordering cost and of the holding cost components, $a + h \sum_{t=i}^{j-1} \tilde{I}_t$. We can therefore associate to arc (i, j) the highest cost that is produced by a current assignment for the closing inventory levels $\tilde{I}_t$, $t \in \{i, \ldots, j - 1\}$, if no variable has been assigned yet, we simply use the minimum possible cost $Q(i, j)$, which we defined before.

4   Experimental Results

In this section we show the effectiveness of our approach by comparing the computational performance of the state-of-the-art CP model with that obtained by our approach. A single problem is considered and the period demands are generated from seasonal data with no trend: $\tilde{d}_t = 50[1 + \sin(\pi t/6)]$. In addition to the "no trend" case (P1) we also consider three others:

(P2) positive trend case, $\tilde{d}_t = 50[1 + \sin(\pi t/6)] + t$
(P3) negative trend case, $\tilde{d}_t = 50[1 + \sin(\pi t/6)] + (52 - t)$
(P4) life-cycle trend case, $\tilde{d}_t = 50[1 + \sin(\pi t/6)] + \min(t, 52 - t)$

In each test we assume an initial null inventory level and a normally distributed demand for every period with a coefficient of variation σ_t/d_t for each $t \in \{1, \ldots, N\}$, where N is the length of the considered planning horizon. We performed tests using four different ordering cost values $a \in \{40, 80, 160, 320\}$ and two different $\sigma_t/d_t \in \{1/3, 1/6\}$. The planning horizon length takes even values in the range $[24, 50]$ when the ordering cost is 40 or 80 and $[14, 24]$ when the

Table 3. Test set P1

		$\sigma_t/d_t = 1/3$								$\sigma_t/d_t = 1/6$							
		$\alpha = 0.95$				$\alpha = 0.99$				$\alpha = 0.95$				$\alpha = 0.99$			
		Filt.		No Filt.		Filt.		No Filt.		Filt.		No Filt.		Filt.		No Filt.	
a	N	Nod	Sec	Nod	Sec	Nod	Sec	Nod	Sec	Nod	Sec	Nod	Sec	Nod	Sec	Nod	Sec
40	40	28	0.4	106	2.9	86	1.2	249	6.4	40	0.6	574	17	10	0.1	192	6.4
	42	28	0.5	95	2.8	87	1.2	233	5.9	40	0.7	582	15	10	0.2	196	5.4
	44	29	0.6	133	4.9	88	1.3	266	8.3	41	0.8	884	26	11	0.2	285	9.0
	46	30	0.8	192	7.8	100	1.9	484	19	44	0.9	3495	120	11	0.2	813	31
	48	39	1.3	444	20	158	3.2	1024	42	66	2.0	5182	190	18	0.5	1208	48
	50	38	0.9	444	21	151	3.6	1024	45	55	1.8	4850	200	15	0.4	1208	52
80	40	52	0.8	1742	78	13	0.2	557	15	19	0.3	9316	300	16	0.3	11276	440
	42	49	0.9	1703	61	13	0.2	530	14	20	0.3	17973	530	17	0.3	22291	690
	44	51	1.0	4810	210	14	0.2	980	26	21	0.4	38751	1400	18	0.4	50805	1600
	46	52	1.1	6063	350	14	0.3	2122	79	31	0.7	103401	4300	18	0.4	111295	4100
	48	57	1.9	20670	1400	15	0.3	5284	210	29	0.7	237112	12000	19	0.5	321998	15000
	50	57	1.7	18938	1300	15	0.3	5284	230	23	0.6	251265	13000	19	0.5	358174	17000
160	14	1	0.0	141	3.0	56	0.2	156	2.5	1	0.0	112	2.6	1	0.0	116	2.4
	16	1	0.0	277	9.0	71	0.3	182	5.1	1	0.0	238	6.7	1	0.0	235	6.8
	18	1	0.0	673	19	50	0.3	393	11	1	0.0	799	24	1	0.0	603	16
	20	1	0.0	3008	82	61	0.5	1359	22	1	0.0	2887	86	1	0.0	2820	75
	22	1	0.0	10620	260	116	1.3	7280	71	1	0.0	14125	380	1	0.0	10739	280
	24	1	0.0	61100	1500	165	1.9	31615	320	1	0.0	70996	1800	1	0.0	59650	1500
320	14	1	0.0	149	4.0	1	0.0	181	4.1	1	0.0	109	3.0	1	0.0	128	3.0
	16	1	0.0	335	12	1	0.0	361	13	1	0.0	246	8.7	1	0.0	284	9.3
	18	1	0.0	813	28	1	0.0	831	28	1	0.0	764	27	1	0.0	700	25
	20	1	0.0	2602	94	1	0.0	2415	82	1	0.0	2114	79	1	0.0	2291	82
	22	1	0.0	7434	260	1	0.0	7416	260	1	0.0	7006	260	1	0.0	6608	230
	24	1	0.0	49663	1600	1	0.0	49299	1500	1	0.0	39723	1400	1	0.0	43520	1500

ordering cost is 160 or 320. The holding cost used in these tests is $h = 1$ per unit per period. Our tests also consider two different service levels $\alpha = 0.95$ ($z_{\alpha=0.95} = 1.645$) and $\alpha = 0.99$ ($z_{\alpha=0.99} = 2.326$). All experiments were performed on an Intel(R) Centrino(TM) CPU 1.50GHz with 500Mb RAM. The solver used for our test is Choco [15], an open-source solver developed in Java. The heuristic used for the selection of the variable is the usual min-domain / max-degree heuristic. The value selection heuristic chooses values in increasing order of size. In our test results a time of 0 means that the Dijkstra algorithm proved optimality at the root node. A header "Filt." means that we are applying our cost-based filtering methods, and "No Filt." means that we solve the instance using only the CP model and the pre-processing methods. Tables 3, 4, 5 and 6 compare the performance of the state-of-the-art CP model with that of our new method.

When a=320, and often when a=160, the Dijkstra algorithm proves optimality at the root node. When $a \in \{40, 80\}$ Dijkstra is unable to prove optimality at the root node, so its main contribution consists in computing lower bounds during the search. However, our method easily solves instances with up to 50 periods, both in term of explored nodes and run time, for every combination of parameters we considered. In contrast, for the CP model both the run times and the number of explored nodes grow exponentially with the number of periods, and the problem becomes intractable for instances of significant size. In all cases our method explores fewer nodes than the pure CP approach, ranging from an improvement of one to several orders of magnitude. Apart from a few trivial

Table 4. Test set P2

		$\sigma_t/d_t = 1/3$								$\sigma_t/d_t = 1/6$							
		$\alpha = 0.95$				$\alpha = 0.99$				$\alpha = 0.95$				$\alpha = 0.99$			
		Filt.		No Filt.		Filt.		No Filt.		Filt.		No Filt.		Filt.		No Filt.	
a	N	Nod	Sec	Nod	Sec	Nod	Sec	Nod	Sec	Nod	Sec	Nod	Sec	Nod	Sec	Nod	Sec
40	40	5	0.1	7	0.1	7	0.1	8	0.1	14	0.3	23	0.4	5	0.1	12	0.2
	42	5	0.1	7	0.1	7	0.1	8	0.1	14	0.2	23	0.4	5	0.1	10	0.1
	44	5	0.1	7	0.1	7	0.1	8	0.1	14	0.3	23	0.5	5	0.1	10	0.2
	46	5	0.1	7	0.1	7	0.1	8	0.2	14	0.3	23	0.5	5	0.1	10	0.2
	48	5	0.1	7	0.2	7	0.1	8	0.2	14	0.3	23	0.5	5	0.1	10	0.2
	50	5	0.1	7	0.2	7	0.2	8	0.2	14	0.3	23	0.6	5	0.1	10	0.2
80	40	24	0.4	4592	14	17	0.3	275	8.3	46	0.9	2565	63	44	0.9	1711	45
	42	24	0.4	4866	13	17	0.3	283	6.7	46	1.0	3027	68	44	0.7	2043	48
	44	24	0.4	5091	15	17	4.7	280	7.9	47	1.1	6024	160	45	0.9	4299	120
	46	46	0.9	5291	45	19	0.4	545	17	51	1.3	14058	410	49	1.1	10311	290
	48	37	0.8	5544	51	19	0.5	545	18	53	1.5	14058	440	53	1.3	10311	310
	50	34	0.7	5850	51	19	0.5	545	19	56	1.8	14079	470	54	1.4	10347	330
160	14	2	0.0	166	3.6	25	0.1	84	1.0	1	0.0	148	2.9	1	0.0	171	3.4
	16	25	0.1	154	4.3	25	0.1	65	1.2	1	0.0	329	8.6	1	0.0	383	11
	18	24	0.1	485	11	27	0.2	174	2.9	1	0.0	948	24	1	0.0	1056	28
	20	34	0.3	2041	35	50	0.4	707	7.9	1	0.0	4228	110	1	0.0	4730	120
	22	50	0.6	9534	120	35	0.3	2954	29	1	0.0	20438	500	1	0.0	23675	530
	24	52	0.5	30502	360	40	0.4	7787	88	1	0.0	71514	1800	1	0.0	83001	1900
320	14	1	0.0	238	5.6	1	0.0	278	6.4	1	0.0	166	3.7	1	0.0	191	4.5
	16	1	0.0	505	17	1	0.0	423	14	1	0.0	387	12	1	0.0	452	14
	18	1	0.0	1447	49	1	0.0	1208	41	1	0.0	1100	34	1	0.0	1268	41
	20	1	0.0	4792	160	1	0.0	4219	150	1	0.0	3992	130	1	0.0	4476	150
	22	1	0.0	20999	670	1	0.0	20417	610	1	0.0	15983	520	1	0.0	18663	600
	24	1	0.0	102158	3200	1	0.0	90398	2600	1	0.0	75546	2500	1	0.0	88602	2800

Table 5. Test set P3

		$\sigma_t/d_t = 1/3$								$\sigma_t/d_t = 1/6$							
		$\alpha = 0.95$				$\alpha = 0.99$				$\alpha = 0.95$				$\alpha = 0.99$			
		Filt.		No Filt.		Filt.		No Filt.		Filt.		No Filt.		Filt.		No Filt.	
a	N	Nod	Sec	Nod	Sec	Nod	Sec	Nod	Sec	Nod	Sec	Nod	Sec	Nod	Sec	Nod	Sec
40	40	3	0.0	5	0.0	3	0.0	4	0.0	7	0.1	9	0.2	3	0.0	5	0.0
	42	3	0.0	5	0.0	3	0.0	4	0.0	7	0.1	9	0.2	3	0.0	5	0.0
	44	4	0.0	7	0.1	4	0.0	6	0.1	8	0.1	14	0.3	4	0.0	7	0.1
	46	9	0.2	15	0.3	5	0.1	13	0.3	17	0.3	40	1.1	8	0.2	14	0.3
	48	8	0.2	15	0.3	5	0.1	13	0.3	17	0.4	56	1.8	14	0.3	25	0.6
	50	7	0.2	15	0.3	5	0.1	13	0.3	17	0.4	56	1.9	13	0.3	25	0.5
80	40	24	0.5	349	10	10	0.1	55	1.2	24	0.4	722	20	10	0.1	310	8.7
	42	26	0.5	354	8.6	8	0.1	53	1.2	23	0.5	1436	35	12	0.2	315	7.5
	44	27	0.5	571	17	9	0.1	88	2.4	24	0.5	3461	110	13	0.2	1053	32
	46	42	1.0	2787	90	10	0.2	258	8.1	37	1.3	10612	360	21	0.5	2881	94
	48	41	1.1	6803	240	10	0.2	385	13	33	1.3	28334	1100	21	0.6	7790	280
	50	42	1.2	6575	250	10	0.2	385	14	35	1.5	26280	1100	21	0.7	7371	280
160	14	7	0.0	23	0.2	9	0.0	16	0.1	14	0.1	53	0.6	10	0.1	29	0.3
	16	5	0.0	19	0.2	9	0.0	18	0.2	19	0.1	52	0.8	9	0.1	26	0.4
	18	7	0.0	42	0.5	10	0.1	30	0.3	21	0.1	149	2.2	11	0.1	87	1.2
	20	17	0.2	137	1.3	12	0.1	70	0.7	23	0.2	512	6.1	20	0.3	310	3.5
	22	9	0.1	376	4.0	15	0.1	221	2.3	28	0.3	1848	18	15	0.2	938	9.4
	24	10	0.2	995	12	25	0.3	543	6.3	37	0.7	4784	55	19	0.2	2471	30
320	14	1	0.0	253	4.2	1	0.0	232	3.8	1	0.0	310	4.4	1	0.0	217	3.4
	16	1	0.0	518	11	1	0.0	518	11	1	0.0	707	14	1	0.0	465	8.5
	18	1	0.0	1475	35	1	0.0	1170	27	1	0.0	1995	44	1	0.0	1416	33
	20	1	0.0	5342	140	1	0.0	4059	96	1	0.0	6678	170	1	0.0	5232	140
	22	1	0.0	21298	550	1	0.0	18065	440	1	0.0	25522	640	1	0.0	21756	560
	24	1	0.0	86072	2300	1	0.0	70969	1800	1	0.0	101937	2800	1	0.0	91358	2400

Table 6. Test set P4

a	N	$\sigma_t/d_t = 1/3$								$\sigma_t/d_t = 1/6$							
		$\alpha = 0.95$				$\alpha = 0.99$				$\alpha = 0.95$				$\alpha = 0.99$			
		Filt.		No Filt.		Filt.		No Filt.		Filt.		No Filt.		Filt.		No Filt.	
		Nod	Sec	Nod	Sec	Nod	Sec	Nod	Sec	Nod	Sec	Nod	Sec	Nod	Sec	Nod	Sec
40	40	7	0.1	21	0.3	11	0.1	24	0.5	30	0.6	89	1.8	7	0.1	33	0.5
	42	7	0.1	18	0.3	11	0.2	21	0.4	30	0.9	91	2.0	7	0.1	31	0.5
	44	8	0.1	32	0.7	12	0.2	37	0.9	31	0.7	152	3.6	8	0.1	51	1.0
	46	14	0.5	83	2.0	14	0.3	93	2.4	46	1.4	474	12.4	13	0.3	126	2.8
	48	12	0.2	83	2.2	14	0.3	93	2.6	56	2.3	735	20.9	19	0.4	188	4.5
	50	11	0.2	83	2.3	14	0.3	93	2.8	58	2.5	735	22.0	18	0.4	188	4.9
80	40	46	0.7	1372	39	24	0.4	433	13	53	1.1	5098	130	36	0.7	2133	55
	42	51	1.5	1673	39	20	0.4	438	11	50	1.1	11452	270	41	1.1	2513	59
	44	52	1.0	2907	74	21	0.4	716	23	52	1.3	27184	780	43	1.3	8776	240
	46	78	2.2	13306	380	23	0.5	2178	74	76	2.4	77332	2600	62	2.3	22582	690
	48	75	1.8	32709	1000	23	0.6	3223	120	76	3.1	202963	7500	61	2.3	60115	2000
	50	77	1.9	31547	1100	23	0.6	3223	130	81	3.2	191836	7600	63	3.3	58171	2100
160	14	11	0.0	166	3.6	25	0.1	84	1.5	1	0.0	148	3.0	1	0.0	171	3.4
	16	9	0.0	154	4.3	25	0.1	65	1.6	1	0.0	329	8.7	1	0.0	383	11
	18	10	0.1	485	11	27	0.1	174	4.0	1	0.0	948	25	1	0.0	1056	28
	20	19	0.2	2041	35	50	0.4	707	12	1	0.0	4228	110	1	0.0	4730	120
	22	17	0.1	9534	120	35	0.3	2954	41	1	0.0	20438	510	1	0.0	23675	540
	24	27	0.4	30502	360	40	0.4	7787	130	1	0.0	71514	1800	1	0.0	83001	1900
320	14	1	0.0	238	5.5	1	0.0	278	8.7	1	0.0	166	3.7	1	0.0	191	4.5
	16	1	0.0	505	17	1	0.0	423	18	1	0.0	387	12	1	0.0	452	14
	18	1	0.0	1447	48	1	0.0	1208	56	1	0.0	1100	34	1	0.0	1268	41
	20	1	0.0	4792	160	1	0.0	4219	200	1	0.0	3992	130	1	0.0	4476	150
	22	1	0.0	20999	660	1	0.0	20417	860	1	0.0	15983	520	1	0.0	18663	600
	24	1	0.0	102158	3200	1	0.0	90398	3700	1	0.0	75546	2700	1	0.0	88602	2800

instances on which both methods take a fraction of a second, this improvement is reflected in the run times.

5 Conclusion

It was previously shown [24] that CP is more natural than mathematical programming for expressing constraints for lot-sizing under the (R^n, S^n) policy, and leads to more efficient solution methods. This paper further improves the efficiency of the CP-based approach by exploiting cost-based filtering. The wide test-bed considered shows the effectiveness of our approach under different parameter configurations and demand trends. The improvement is several orders of magnitude in almost every instance we analyzed. We are now able to solve to optimality problems of a realistic size with planning horizons of fifty and more periods, in times of less than a second and often without search, since the bounds produced by our DP relaxation proved to be very tight in many instances. In future work we aim to extend our model to new features such as lead-time for orders and capacity constraints for the inventory.

Acknowledgements. This work was supported by Science Foundation Ireland under Grant No. 03/CE3/I405 as part of the Centre for Telecommunications Value-Chain-Driven Research (CTVR) and Grant No. 00/PI.1/C075.

References

1. Apt, K.: Principles of Constraint Programming. Cambridge University Press, Cambridge, UK (2003)
2. Askin, R.G.: A Procedure for Production Lot Sizing With Probabilistic Dynamic Demand. AIIE Transactions 13, 132–137 (1981)
3. Bellman, R.E.: Dynamic Programming. Princeton University Press, Princeton, NJ (1957)
4. Florian, M., Lenstra, J.K., Rinooy, A.H.G.: Deterministic Production Planning: Algorithms and Complexity. Management Science 26(7), 669–679 (1980)
5. Bookbinder, J.H., Tan, J.Y.: Strategies for the Probabilistic Lot-Sizing Problem With Service-Level Constraints. Management Science 34, 1096–1108 (1988)
6. Brailsford, S.C., Potts, C.N., Smith, B.M.: Constraint Satisfaction Problems: Algorithms and Applications. European Journal of Operational Research 119, 557–581 (1999)
7. Fahle, T., Sellmann, M.: Cost-Based Filtering for the Constrained Knapsack Problem. Annals of Operations Research 115, 73–93 (2002)
8. Focacci, F., Lodi, A., Milano, M.: Cost-Based Domain Filtering. In: Jaffar, J. (ed.) Principles and Practice of Constraint Programming – CP'99. LNCS, vol. 1713, pp. 189–203. Springer, Heidelberg (1999)
9. Fortuin, L.: Five Popular Probability Density Functions: a Comparison in the Field of Stock-Control Models. Journal of the Operational Research Society 31(10), 937–942 (1980)
10. Gartska, S.J., Wets, R.J.-B.: On Decision Rules in Stochastic Programming. Mathematical Programming 7, 117–143 (1974)
11. Gupta, S.K., Sengupta, J.K.: Decision Rules in Production Planning Under Chance-Constrained Sales. Decision Sciences 8, 521–533 (1977)
12. Heady, R.B., Zhu, Z.: An Improved Implementation of the Wagner-Whitin Algorithm. Production and Operations Management 3(1) (1994)
13. Johnson, L.A., Montgomery, D.C.: Operations Research in Production Planning, Scheduling, and Inventory Control. Wiley, New York (1974)
14. Jussien, N., Debruyne, R., Boizumault, P.: Maintaining Arc-Consistency Within Dynamic Backtracking. In: Dechter, R. (ed.) CP 2000. LNCS, vol. 1894, pp. 249–261. Springer, Heidelberg (2000)
15. Laburthe, F., OCRE project team.: Choco: Implementing a CP Kernel. Bouygues e-Lab, France
16. Lustig, I.J., Puget, J.-F.: Program Does Not Equal Program: Constraint Programming and its Relationship to Mathematical Programming. Interfaces 31, 29–53 (2001)
17. Peterson, R., Silver, E., Pyke, D.F.: Inventory Management and Production Planning and Scheduling. John Wiley and Sons, New York (1998)
18. Porteus, E.L.: Foundations of Stochastic Inventory Theory. Stanford University Press, Stanford, CA (2002)
19. Sedgewick, R.: Algorithms. Addison-Wesley Publishing Company, Reading, Massachusetts (1983)
20. Silver, E.A.: Inventory Control Under a Probabilistic Time-Varying Demand Pattern. AIIE Transactions 10, 371–379 (1978)
21. Simchi-Levi, D., Simchi-Levi, E., Kaminsky, P.: Designing and Managing the Supply Chain. McGraw-Hill, New York (2000)

22. Tan, J.Y.: Heuristic for the Deterministic and Probabilistic Lot-Sizing Problems. Department of Management Sciences, University of Waterloo (1983)
23. Tarim, S.A., Kingsman, B.G.: The Stochastic Dynamic Production/Inventory Lot-Sizing Problem With Service-Level Constraints. International Journal of Production Economics 88, 105–119 (2004)
24. Tarim, S.A., Smith, B.: Constraint Programming for Computing Non-Stationary (R,S) Inventory Policies. European Journal of Operational Research (to appear)
25. Tarim, S.A., Manandhar, S., Walsh, T.: Stochastic Constraint Programming: A Scenario-Based Approach. Constraints 11, 53–80 (2006)
26. Vajda, S.: Probabilistic Programming. Academic Press, New York (1972)
27. Wagner, H.M., Whitin, T.M.: Dynamic Version of the Economic Lot Size Model. Management Science 5, 89–96 (1958)
28. Zipkin, P.H.: Foundations of Inventory Management. McGraw-Hill/Irwin, Boston, Mass (2000)
29. Silver, E.A., Pyke, D.F., Peterson, R.: Inventory Management and Production Planning and Scheduling. John-Wiley and Sons, New York (1998)
30. Wemmerlov, U.: The Ubiquitous EOQ - Its Relation to Discrete Lot Sizing Heuristics. Internat. J. of Operations and Production Management 1, 161–179 (1981)
31. Van Hentenryck, P., Carillon, J.-P.: Generality vs. specificity: an experience with AI and OR techniques. In: National Conference on Artificial Intelligence (AAAI-88) (1988)
32. Charnes, A., Cooper, W.W.: Chance-Constrainted Programming. Management Science 6(1), 73–79 (1959)
33. Birge, J.R., Louveaux, F.: Introduction to Stochastic Programming. Springer, New York (1997)

Author Index

Lecture Notes in Artificial Intelligence (LNAI)

Vol. 4384: T. Washio, K. Satoh, H. Takeda, A. Inokuchi (Eds.), New Frontiers in Artificial Intelligence. IX, 401 pages. 2007.

Vol. 4371: K. Inoue, K. Satoh, F. Toni (Eds.), Computational Logic in Multi-Agent Systems. X, 315 pages. 2007.

Vol. 4369: M. Umeda, A. Wolf, O. Bartenstein, U. Geske, D. Seipel, O. Takata (Eds.), Declarative Programming for Knowledge Management. X, 229 pages. 2006.

Vol. 4342: H. de Swart, E. Orłowska, G. Schmidt, M. Roubens (Eds.), Theory and Applications of Relational Structures as Knowledge Instruments II. X, 373 pages. 2006.

Vol. 4335: S.A. Brueckner, S. Hassas, M. Jelasity, D. Yamins (Eds.), Engineering Self-Organising Systems. XII, 212 pages. 2007.

Vol. 4334: B. Beckert, R. Hähnle, P.H. Schmitt (Eds.), Verification of Object-Oriented Software. XXIX, 658 pages. 2007.

Vol. 4333: U. Reimer, D. Karagiannis (Eds.), Practical Aspects of Knowledge Management. XII, 338 pages. 2006.

Vol. 4327: M. Baldoni, U. Endriss (Eds.), Declarative Agent Languages and Technologies IV. VIII, 257 pages. 2006.

Vol. 4314: C. Freksa, M. Kohlhase, K. Schill (Eds.), KI 2006: Advances in Artificial Intelligence. XII, 458 pages. 2007.

Vol. 4304: A. Sattar, B.-h. Kang (Eds.), AI 2006: Advances in Artificial Intelligence. XXVII, 1303 pages. 2006.

Vol. 4303: A. Hoffmann, B.-h. Kang, D. Richards, S. Tsumoto (Eds.), Advances in Knowledge Acquisition and Management. XI, 259 pages. 2006.

Vol. 4293: A. Gelbukh, C.A. Reyes-Garcia (Eds.), MICAI 2006: Advances in Artificial Intelligence. XXVIII, 1232 pages. 2006.

Vol. 4289: M. Ackermann, B. Berendt, M. Grobelnik, A. Hotho, D. Mladenič, G. Semeraro, M. Spiliopoulou, G. Stumme, V. Svátek, M. van Someren (Eds.), Semantics, Web and Mining. X, 197 pages. 2006.

Vol. 4285: Y. Matsumoto, R.W. Sproat, K.-F. Wong, M. Zhang (Eds.), Computer Processing of Oriental Languages. XVII, 544 pages. 2006.

Vol. 4274: Q. Huo, B. Ma, E.-S. Chng, H. Li (Eds.), Chinese Spoken Language Processing. XXIV, 805 pages. 2006.

Vol. 4265: L. Todorovski, N. Lavrač, K.P. Jantke (Eds.), Discovery Science. XIV, 384 pages. 2006.

Vol. 4264: J.L. Balcázar, P.M. Long, F. Stephan (Eds.), Algorithmic Learning Theory. XIII, 393 pages. 2006.

Vol. 4259: S. Greco, Y. Hata, S. Hirano, M. Inuiguchi, S. Miyamoto, H.S. Nguyen, R. Słowiński (Eds.), Rough Sets and Current Trends in Computing. XXII, 951 pages. 2006.

Vol. 4253: B. Gabrys, R.J. Howlett, L.C. Jain (Eds.), Knowledge-Based Intelligent Information and Engineering Systems, Part III. XXXII, 1301 pages. 2006.

Vol. 4252: B. Gabrys, R.J. Howlett, L.C. Jain (Eds.), Knowledge-Based Intelligent Information and Engineering Systems, Part II. XXXIII, 1335 pages. 2006.

Vol. 4251: B. Gabrys, R.J. Howlett, L.C. Jain (Eds.), Knowledge-Based Intelligent Information and Engineering Systems, Part I. LXVI, 1297 pages. 2006.

Vol. 4248: S. Staab, V. Svátek (Eds.), Managing Knowledge in a World of Networks. XIV, 400 pages. 2006.

Vol. 4246: M. Hermann, A. Voronkov (Eds.), Logic for Programming, Artificial Intelligence, and Reasoning. XIII, 588 pages. 2006.

Vol. 4223: L. Wang, L. Jiao, G. Shi, X. Li, J. Liu (Eds.), Fuzzy Systems and Knowledge Discovery. XXVIII, 1335 pages. 2006.

Vol. 4213: J. Fürnkranz, T. Scheffer, M. Spiliopoulou (Eds.), Knowledge Discovery in Databases: PKDD 2006. XXII, 660 pages. 2006.

Vol. 4212: J. Fürnkranz, T. Scheffer, M. Spiliopoulou (Eds.), Machine Learning: ECML 2006. XXIII, 851 pages. 2006.

Vol. 4211: P. Vogt, Y. Sugita, E. Tuci, C.L. Nehaniv (Eds.), Symbol Grounding and Beyond. VIII, 237 pages. 2006.

Vol. 4203: F. Esposito, Z.W. Raś, D. Malerba, G. Semeraro (Eds.), Foundations of Intelligent Systems. XVIII, 767 pages. 2006.

Vol. 4201: Y. Sakakibara, S. Kobayashi, K. Sato, T. Nishino, E. Tomita (Eds.), Grammatical Inference: Algorithms and Applications. XII, 359 pages. 2006.

Vol. 4200: I.F.C. Smith (Ed.), Intelligent Computing in Engineering and Architecture. XIII, 692 pages. 2006.

Vol. 4198: O. Nasraoui, O. Zaïane, M. Spiliopoulou, B. Mobasher, B. Masand, P.S. Yu (Eds.), Advances in Web Mining and Web Usage Analysis. IX, 177 pages. 2006.

Vol. 4196: K. Fischer, I.J. Timm, E. André, N. Zhong (Eds.), Multiagent System Technologies. X, 185 pages. 2006.

Vol. 4188: P. Sojka, I. Kopeček, K. Pala (Eds.), Text, Speech and Dialogue. XV, 721 pages. 2006.

Vol. 4183: J. Euzenat, J. Domingue (Eds.), Artificial Intelligence: Methodology, Systems, and Applications. XIII, 291 pages. 2006.

Vol. 4180: M. Kohlhase, OMDoc – An Open Markup Format for Mathematical Documents [version 1.2]. XIX, 428 pages. 2006.

Vol. 4177: R. Marín, E. Onaindía, A. Bugarín, J. Santos (Eds.), Current Topics in Artificial Intelligence. XV, 482 pages. 2006.

Vol. 4160: M. Fisher, W. van der Hoek, B. Konev, A. Lisitsa (Eds.), Logics in Artificial Intelligence. XII, 516 pages. 2006.

Vol. 4155: O. Stock, M. Schaerf (Eds.), Reasoning, Action and Interaction in AI Theories and Systems. XVIII, 343 pages. 2006.

Vol. 4149: M. Klusch, M. Rovatsos, T.R. Payne (Eds.), Cooperative Information Agents X. XII, 477 pages. 2006.